MAKER OF PATTERNS

ALSO BY FREEMAN DYSON

Dreams of Earth and Sky

A Many-Colored Glass

The Scientist as Rebel

Imagined Worlds

From Eros to Gaia

Infinite in All Directions

Origins of Life

Weapons and Hope

Disturbing the Universe

My Dear Family

I am afraid I have frittered a great deal of time lately, so that I now find myself very busy and I have to get this letter off this evening.

I have had a session or two in the bookshops, and have spent a lot of money; but all such good value that I cannot help it. I hope that if I continue at this rate I shall be able to pay my bills, but I expect I shall.

I am not, as Alice seemed to insinuate, quite as badly fed as you imagine. I have been eating a great deal of brown bread; and as I have to get through the loaf before it is stale, the reverse of starvation sometimes overtakes me, i.e. a tendency to sleep long hours.

I look forward to seeing you on Monday; it is a pity you chose Tuesday which is my busiest day, but it is not very busy by any means.

I had a very good tea-party on Sunday with the Edwards'. It was a large party but very amusing, and I no longer despise tea-parties so much as I did when I had to spend 3/4 of the time in such idle pursuits. Mr. Edwards himself looked very worn, but talked in fine style.

week later you on to Cambridge. — pick up some boots, I should be back and visit you About a for a few days.

Much Love   Freeman.

MAKER OF PATTERNS

AN AUTOBIOGRAPHY THROUGH LETTERS

FREEMAN DYSON

LIVERIGHT PUBLISHING CORPORATION

A DIVISION OF W. W. NORTON & COMPANY

INDEPENDENT PUBLISHERS SINCE 1923

New York · London

For information about permission to reproduce selections from this book,
write to Permissions, Liveright Publishing Corporation, a division of
W. W. Norton & Company, Inc., 500 Fifth Avenue, New York, NY 10110

For information about special discounts for bulk purchases, please contact
W. W. Norton Special Sales at specialsales@wwnorton.com or 800-233-4830

Manufacturing by Quad Graphics, Fairfield
Book design by Marysarah Quinn
Production manager: Beth Steidle

ISBN 978-0-87140-386-5

Liveright Publishing Corporation, 500 Fifth Avenue, New York, N.Y. 10110
www.wwnorton.com

W. W. Norton & Company Ltd., 15 Carlisle Street, London W1D 3BS

1 2 3 4 5 6 7 8 9 0

A mathematician, like a painter or a poet, is a maker of patterns. If his patterns are more permanent than theirs, it is because they are made with ideas.

—GODFREY HAROLD HARDY

CONTENTS

PREFACE

IN MARCH 2017, when this book was almost finished, my wife received a message from our twelve-year-old granddaughter: "We are all metaphors in this dark and lonely world." Our daughter added her own comment, "The sentiment is tempered by the fact that she has a pink Afro." The pink Afro displays a proud and joyful spirit, masking the melancholy thoughts of a teenager confronting an uncertain future. Our granddaughter is now emerging into a world strikingly similar to the world of 1936 into which I came as a twelve-year-old. Both our worlds were struggling with gross economic inequality, stubbornly persistent poverty, brutal dictators on the rise, and small wars presaging worse horrors to come. I too was a metaphor for a new generation of young people without illusions. Her declaration of independence is a pink Afro. Mine was a passionate pursuit of mathematics. I escaped from the barbaric world of Hitler and Stalin into the abstract world of Hardy. Hardy was the most famous mathematician in England when I came to him as a student in 1941. He taught me to be a maker of patterns. An even greater maker of patterns was the Indian genius Srinivasa Ramanujan, who had come to Hardy as a student in 1914 and died at age thirty-two in 1920. Through Hardy, I entered Ramanujan's magical world. My first discoveries were concerned with the numbers 5 and 7 which play special roles in the weird arithmetical patterns of Ramanujan.

Afterwards I found patterns of comparable beauty in the dance of electrons jumping around atoms. Patterns as elegant as those of Ramanujan had been discovered in the world of physics by Paul Dirac, whose lectures on quantum mechanics I also attended. Dirac was one of the pioneers who had entered the strange new world of quantum physics, where strict causality is abandoned and atomic events occur by chance. The idea that chance governs nature was then still open to question. In the world of human affairs, Lev Tolstoy asked the same question, whether free choice prevails. While Dirac proclaimed free choice in the world of physics, Tolstoy denied it in the world of history. The idea that Dirac called causality, Tolstoy called Providence. At the end of his *War and Peace*, he wrote a long philosophical discussion, explaining why human free will is an illusion and Providence is the driving force of history. When I was a student in Cambridge, the same Providence that had destroyed Napoleon's army in Russia in 1812 was destroying Hitler's army in Russia in 1943. I was reading Tolstoy and Dirac at the same time.

The words *maker of patterns* had for Hardy a double meaning. Patterns could be made with words as well as with ideas, with books as well as with theorems. I once asked him why in his old age, when he was sixty-five and I was eighteen, he had stopped exploring new mathematics and was spending his time writing books. He answered without hesitation, "Young men should prove theorems: old men should write books." He took pride in his style as a writer. He was a maker of patterns in English prose as well as in the theory of numbers. In my later life I followed his example. As a young man I wrote technical papers, exploring patterns of ideas in mathematics and physics. As an old man I write books, exploring patterns of words in literature and history.

The letters collected in this book record a cycle in the history of the world from 1936 to 1978, from the civil war in Spain to the rise of Gorbachev in the Soviet Union. This was a cycle rolling from doom and gloom in the 1930s, to death and disaster in the 1940s, to fear and trembling in the 1950s, to smaller disasters in the 1960s, to recovery and promise in the 1970s. Contrasting with this cycle of death and rebirth in the political world, there was steady progress and a continued succession of major advances in the world of science. Two discoveries transformed our views of the biological and physical worlds. The discovery of the double helix structure of the DNA molecule by Francis Crick and Jim Watson in 1953 made the basic processes of life suddenly accessible to study with the tools of physics and chemistry. The mysteries of biological reproduction and heredity could be translated into testable molecular models. The discovery of the cosmic background radiation by Arno Penzias and Robert Wilson in 1964 opened the entire physical universe to our observation. Suddenly the whole universe back to its beginnings was visible to our instruments.

In 1968 I read *The Double Helix*, the story written by Watson describing how he and Crick discovered the secret of life. The secret of life, as they proudly boasted after they made the discovery, was the structure of the DNA molecules that carry the genetic information in every cell of every living creature larger than a virus. The helical structure, with the genetic information written backward and forward in two complementary strands around the helix, gives the molecules the ability to replicate themselves precisely when the cell divides. The two daughter cells are born with exact copies of the mother's genetic information. Watson's book tells in vivid detail how the discovery happened, not by logical scientific reasoning but in a personal drama with real human characters. He brings the characters to life with verbatim accounts of their conversations, stumbling and squabbling as they grope their way toward the truth.

Soon after I read Watson's book, I met him in person. I asked him how he could possibly remember the details of the conversations and arguments that he put into his story. He answered, "Oh, that was easy. I wrote a letter every week to my mother in America describing my life in England, and she kept the letters." I had been writing a letter every week to my parents in England, describing my life in America. That same day I wrote to my mother, urging her to keep the letters. She kept them, giving me the raw material for this book.

I do not have any great discovery like the double helix to describe. The letters record the daily life of an ordinary scientist doing ordinary work. I find them interesting because I had the good fortune to live through extraordinary historical times with an extraordinary collection of friends. Letters are valuable witnesses to history because they are written without hindsight.

They describe events as they appeared to the participants at the time. Later memories of the same events may be seriously distorted by hindsight. When I compare my memories with the letters, I see that I not only forget things, I also remember things that never happened.

A striking example of a false memory was a lunch party in 1958 at the home of Robert Oppenheimer and his wife Kitty. Two other couples were there, the theologian Reinhold Niebuhr and his wife Ursula, who had come to the Institute for Advanced Study for a year, and the retired diplomat George Kennan and his wife Annelise, who had come permanently so that George could pursue his second career as a historian. After lunch we sat around a glowing wood fire in the living room. Oppenheimer pulled a beautiful leather-bound volume from his bookcase and read a poem, "The Pulley" by George Herbert (1633), in a beautifully cadenced voice. George Kennan had not known Herbert's poetry before. Oppenheimer said it was high time for the two Georges to get to know each other. Then he turned toward Ursula Niebuhr and said, "But you of course know him well." It turned out that George Herbert was a distant ancestor of Ursula. Oppenheimer then continued to read other poems of Herbert, and so the afternoon continued with personal warmth and poetry around the fireside. I have a vivid memory of sitting with that group around the fireside. Recently I was reading some old letters and found one written long ago by Ursula Niebuhr, describing that lunch party and confirming my memory. Every detail of my memory is correct except for one. I was not there. My memory has somehow stolen that scene from Ursula and put me into it.

All through a long life I had three main concerns, with a clear order of priority. Family came first, friends second, and work third. The same order of priority appears in the letters. The best passages describe the human story of two marriages and growing children. The next best describe the community of friends that I enjoyed as colleagues in science and in public affairs. The details of my work as a scientist are barely mentioned. The neglect of science is mainly due to the fact that my parents were not interested in technical details. I assume that readers of this book will also not be interested in technical details.

I wrote the letters to my family as a dutiful son, whenever I was separated from them, beginning in 1941, when I went to Cambridge University as a student, and continuing intermittently until the death of my sister Alice in 2012. The family consisted of my father, Sir George, who died in 1964; my mother, Mildred, who died in 1975; and my sister, who lived independently in London for many years and moved to the parents' home in Winchester to take care of our mother after our father died. All three wrote back to me frequently. I had originally intended to include in this book the letters of all four of us, giving the reader a picture of life as it was lived on both sides of the Atlantic Ocean. Reluctantly I decided to include only my own letters. To include the incoming letters would have doubled the size of the book and would have made the narrative less coherent. Readers who are interested in the life and work of my father should go to the excellent biography by Paul Spicer (2014).

When the letters begin in 1941, World War II is raging in Russia. Hitler's armies, after defeating France and occupying most of Europe, are marching deep into the Soviet Union. Britain is enjoying a peaceful interlude between the disasters in France in 1940 and in Singapore in 1942. My parents are bombed out of their house in London and have found refuge with a friendly family in Reading. My father is director of the Royal College of Music in London and is keeping the college open, having moved a bed into his office so that he can stay overnight during the bombing. The trustees of the college had wanted to close it, but he was determined to keep it open. He told them that he intended to stay there so long as there was a roof over his head. The trustees gave way and agreed to keep it open.

Defying Hitler in 1940 was for my father, as it was for Winston Churchill, his finest hour. By keeping the college open, he set a good example for several other institutions that had made plans to evacuate but decided to stay open, notably the Sadlers Wells Opera and the Old Vic Theatre. All of them stayed active throughout the war and kept London alive as a cultural center. Paul Spicer in his biography quotes from a letter written by Hazel Bole, who was a student at the college in 1940: "I was fire-watching on the roof during the air raids. We students were there to throw sand on the incendiary bombs which the German bombers were raining down on us. One night I grabbed the large bucket, and someone else grabbed it too. I let go, and in a sudden flash of fireball I saw Sir George grinning at me." Later in the war, my parents moved into an apartment close to the Royal College, and my sister moved to St. Thomas's Hospital in London for training as a medical social worker. My father retired in 1952, and my parents then moved back to Winchester, where we had lived until 1938, when he became director.

The letters selected for this book cover only the period 1941–78, the first half of my adult life. The letters continue for thirty more years, but the second half of a life is usually less interesting than the first half. I believe it was Rudyard Kipling who said that a man should get half of his dying done before the age of forty. I decided to stop the selection at 1978, to keep the book short and readable while including the high points of my story. Throughout the book except in this preface, the texts of letters are printed in Roman font, and my later comments are printed in italics.

I am grateful to Robert Weil, my editor at Norton, for his unfailing help and encouragement. I am grateful to my son George and to Kimberly Jacobsen for digitizing the letters and making them easy to read and search. Finally, I am grateful to my wife Imme for the tender loving care that keeps me alive.

GREAT MINDS AROUND
THE BILLIARD TABLE

THE LETTERS BEGAN *when I arrived as a seventeen-year-old undergraduate at Trinity College, Cambridge, in September 1941. It was a great time to get an education, in the middle of World War II. The famous old professors were all there, but there were hardly any students. As a dutiful son, I wrote frequent letters to my parents. In this selection of the letters, nothing has been added, but a great deal has been silently subtracted. Here is a list of the main characters who appear in Chapter 1.*

> *Godfrey Harold Hardy, then aged sixty-four, the most famous pure mathematician in England, known to the public for his book,* A Mathematician's Apology *(1940). The book is short and beautifully written. I quote in the epigraph two sentences from his book, from which I also borrowed the title for mine.*
>
> *John Edensor Littlewood, then aged fifty-six, had worked with Hardy in a famous collaboration for many years, exploring the theory of numbers. Some years earlier they decided that they had done enough with number theory and made a pact to work in the future on other subjects. Littlewood never married, but he had a daughter. When I knew him in Cambridge, the daughter was only a rumor. Many years later the daughter finally emerged as a real person.*

Abram Samoilovich Besicovitch, then aged fifty, was a Russian mathematician who stayed in Russia after the revolution and settled in Cambridge ten years later. He carried the ancient geometry of Euclid to a deeper level, discovering new and rich structures in the menagerie of sets of points in a Euclidean plane. He was separated from his wife, but she was also living in Cambridge, and they remained friends. Hardy, Littlewood, and Besicovitch were all living in rooms at Trinity College. Because of the war, there were no young mathematicians in Cambridge. The three who gathered around the billiard table all belonged to the nineteenth century. In Cambridge we learned nineteenth-century mathematics. We knew little of the twentieth-century mathematics that had grown up in the 1930s in France, with a far more abstract style and a new set of general concepts. Only later when the war was over, French mathematics swept over England and America, and our elderly heroes became quaint and old-fashioned.

Yelizaveta Fyodorovna Hill, then aged forty, was head of the department of Slavic languages. She had grown up in St. Petersburg with a British father and a Russian mother. They emigrated to England in 1918.

Prince Dimitri Obolensky, then aged twenty-three and a student at Trinity College, was later a distinguished scholar and professor of Russian history at Oxford. Born in Russia in 1918, he escaped to England as a baby with his parents.

Paul Maurice Adrien Dirac, then aged thirty-nine, was the leading British physicist who stayed in Cambridge during the war.

The billiard table was for two years the center of my life. It belonged to Besicovitch and stood in the big room where he entertained guests on the ground floor of his home in Neville's Court. He was addicted to billiards, and Hardy, who had been a passionate cricketer in his youth, was a passionate billiard player in his old age. Whenever I needed to talk with either of them, I could usually find them at the billiard table. Fortunately my father had bought a billiard table when I was a child and taught me how to play. Our table was more modest than Besicovitch's, but I could play well enough to pass the time with the famous professors.

My intense interest in Russia began when I was in high school and became even stronger when Hitler invaded in 1941 and the Russians became our heroic allies. I planned to go to the Soviet Union after the war to study the language and literature and work with Russian mathematicians and scientists. When the war ended, Stalin made it clear that foreign students were unwelcome, and I switched my travel plans to the United States.

I wrote all the letters in this chapter to my parents from Trinity College, except for the two from the climbing hut in Wales.

OCTOBER 19, 1941

I have now consolidated my position and feel that I can begin doing things because I want to and not because I have to. I have joined two mathematical societies which hold meetings occasionally. The first meeting was a lecture to the "Archimedeans" on "The Place of Mathematics in a Planned Society" by Professor Bernal, the perpetrator of "The Social Function of Science." The trouble with him, as I knew it would be, was that he knows nothing about mathematics, even in its most applied forms. The only good thing about it was that he did proclaim that there is something for mathematicians to do in the statistical side of things; but he did not say anything definite about it. Hardy was there and adopted a very tolerant attitude; he seems to have been more impressed than I was. I have also joined Trinity College Orchestra, which holds its first meeting today. I have so far only looked at my instrument once, when I practiced for an hour last night to see if I could still play. I found I could, and perhaps the orchestra will make me practice. If not, I will have to see about having lessons.

My instrument was a violin, which I learned to play as a child. My father was wise enough to see that I had no real musical talent and never put pressure on me to play well. I played barely well enough to be acceptable in a student orchestra.

I have joined the Cambridge University Mountaineering Club which may prove a fateful decision, as it has meets in every vacation lasting a week, and I shall go to most of them. It has lectures by Great Men periodically; the first is tomorrow. At the meets there is a large hut where the people sleep; and parties of various degrees of competence go out every day to do their worst. It is, they say, the leading climbing club in the country as far as really difficult climbing is concerned; but they assure me it caters for novices in large numbers. I shall have to get boots, ropes, and all the other lethal weapons. The Christmas meet is in North Wales every year. In peacetime they went abroad to Alps and Pyrenees. I have not yet succeeded in making any new friends. If ever I do need friends, I have no doubt I shall find some on the precipices of Snowdon.

I have fixed up all my lectures now, they are: Hardy on Fourier series, Besicovitch on integration, Dirac on quantum mechanics, Pars on dynamics. The lectures are very select; Hardy has an audience of four, Besicovitch three, Pars four, and Dirac about twenty. All these lectures are, at any rate in parts, new to me; and all the lecturers know how to lecture; so I enjoy them very much. Dirac is very slow and easy to follow; Pars and Besicovitch a bit quicker, but still comfortable; Hardy goes like an avalanche and it is all I can do to keep up with him. One learns about three times as much from Hardy in an hour as from anyone else; it is a testing business keeping the thread of his arguments. If one can do it, one can do almost anything in analysis. Fortunately, though I have not had much practice, I know the tricks of the trade and am seldom entirely fogged.

I went to the university library on Monday. It is a vast place, with every book ever printed in it somewhere (with no doubt a few exceptions). I browsed in its labyrinths for a few hours and found a lot of interesting stuff. The best part of it, however, was the tea room which gives the best value for money I have seen for a long time. On the whole I shall not go there much except to have tea, for all the books one could ever read are in the Trinity Library. This is a very comfortable place, with any amount of journals in many languages. If ever I start investigating some particular subject, I shall find all I want there.

I find there is no compulsion, or even suggestion, for me to do anything in the way of duty, military or civil, until I have to register (sometime next year). In any case the air squadron would not take me till my last year. There do not seem to be any intellectual societies like the Winchester ones which were so ubiquitous. All the societies are either purely political or purely technical; general education seems to be entirely neglected. However, I have had enough to last me a bit.

Winchester College was the high school that I attended for five years before going to Cambridge. It was an ancient school, founded in 1382 to educate the young Catholic priests who became the permanent bureaucracy running England while the kings were away fighting in France. The school survived the Protestant Reformation and maintained high intellectual standards. The boys enjoyed great freedom, with plenty of time outside the classroom to pursue their own interests. Informal clubs proliferated. I belonged to the Obscure Languages Club, whose main purpose was to speak obscenities in as many languages as possible. My father had been a teacher at the same school, running the music department, playing the organ in chapel services, and conducting choirs and orchestras. He became well known as a composer and broadcaster. He left the school in 1938 to become director of the Royal College of Music.

Both at Winchester College and at the Royal College, the war that began for England in 1939 hit us hard. We knew that we were in it for the long haul, with no end in sight, with a high probability that little that we valued would survive. And yet in a paradoxical way, our response to the war in both places was to ignore it as far as possible. My father in London, and our teachers at Winchester, understood that the best way to show our contempt for Hitler was to continue making music and to continue studying Latin and Greek, as if Hitler did not exist. My father said to the students in London in 1940, "All we have to do is to behave halfway decently, and the whole world will come to our side." That was his way of fighting Hitler. The paradoxical result is that in these letters written during the darkest years of the war, the war is hardly mentioned. In Cambridge, just as in London and in Winchester, the way to defend England was to make sure that there would be something in England worth defending.

OCTOBER 25, 1941

On Monday I went to the mountaineering club. Dr. Jacub Bahar on "The Tatra and the Himalaya" with slides. He is a Pole who climbed Nanda Devi East Peak in 1939. Of a party of four, the leader and another were killed on another mountain afterwards. This man told the story beautifully, and the slides were magnificent. The Tatra are on the Polish-Slovakian frontier and are, apparently, very climbable. They are the chief mountaineering resort of Eastern Europe. My lectures are all proceeding according to plan. I have discovered more about analysis in eight hours of Hardy's "Fourier series" than could be derived from many volumes. And even old Besicovitch hits the nail on the head occasionally. Pars is making dynamics a great deal less repellent; in fact it is worth doing if you do it like him. Dirac is the only lecturer who does not break new ground as far as I am concerned, but he suffers from an audience of twenty. Later on when he stops dealing with the fundamentals, which are mathematical, which I know and they don't, he will go on to physics, which I don't know and they do.

I was unexpectedly one morning appointed "stair-case marshal" which means that I have to look after my staircase, put out bombs and carry out corpses, if a bomb happens to burst within twenty yards of us. All my duties have amounted to so far is trying to get a stirrup-pump mended by plumbers who know nothing about it.

The stirrup-pump was standard equipment during the war for putting out fire-bombs. It stood in a bucket filled with water and pumped the water into a hose that was directed at the fire. It was hand-operated and needed no electricity. I never had a chance to use it.

NOVEMBER 2, 1941

I have started on some work of my own. On Thursday Besicovitch gave me a medium-size book entitled *L'Hypothèse du Continu* by Waclaw Sierpinski, the great Polish mathematician, and told me to come back when I had mastered chapters 1 and 2. I having read them (seventy-five pages merely) with understanding except for fairly unimportant technical terms, I went to see him on Saturday; and after elucidating certain points he propounded a problem for me to work on. He gave me several papers to read, by himself and Sierpinski chiefly; one of them is a very elegant proof of an allied problem found by Besicovitch in the last two months. All the papers refer to Besicovitch as quite a great man; they talk about "Besicovitch measure," "Besicovitch dimension," and so on. I have not yet got to grips with my problem; I must read the papers before I know what it means. I shall have an interesting time whether it proves amenable or not. Besicovitch says it will be very good if I solve it in a few months.

SUNDAY, BETTWS-Y-COED HELYG FARM

I am safely installed here after a very exhausting journey. The farm consists of two rooms, one for sleeping and drying the clothes and the other for everything else. The table seats ten at a pinch; it will be even more intimate when we are twenty. It is a great triumph having got here. These Welsh are very suspicious. For the first time in my life I was asked for my identity card, by a body of miscellaneous women, and then again later by a policeman. At six-thirty a.m. in Bangor I met an unfortunate trombone player who was trying to find the BBC but did not know where he was or where they were. I hope he is not still roaming the streets with his trombone. I was more fortunate. It took six hours to cover the twelve miles from Bangor to here. Never again will I take a suitcase on foot for long distances, though actually I got lifts for about half the distance.

The trombonist was probably looking for the BBC studio where the Welsh National Radio programs were broadcast.

FRIDAY, HELYG

My career has suffered a temporary eclipse after a most promising start; yesterday I went out with two gallant gentlemen to climb a cliff known as something Ddu, which proved rather difficult as it was encased in about six inches of moss. The first gallant, by name Bumpstead, who is an old stager, led; and progressed upwards with immense deliberation, stopping now and then to restore the circulation to his fingers. He complained that a river was flowing down his sleeves and out at his boots. We two sat in the rain at the bottom for about an hour, after which the second man started. I sat disconsolately underneath paying out the rope inch by inch. At last when there were only about five feet of rope left there was a vague shout from above. A little later I felt a sharp blow on the side of the head. Feeling very cold and tired, I retired to a secluded corner and wept silently; then all of a sudden I seemed to be soused in blood from head to foot, and I must confess ignoble joy filled my breast with the thought of an early return to the warmth of Helyg. I shouted to Bumpstead to come to my assistance. Fortunately he was able to come to the bottom quickly, after the second man had scrambled to the top. It was his first careless rapture on reaching the top that caused the stone to fall.

Bumpstead efficiently bound me up with my two handkerchiefs, which stopped the bleeding; and we set off leisurely downhill to Helyg, where I was given a pint of tea and felt very happy. Finally a doctor arrived and spent a half-hour poking about with razors, probes, and needles, which was about as bad as an average session with a ruthless dentist. He finished by sewing it up with silk. He said that it was a small cut but happened to have punctured an artery and that nothing effective could be done without opening up the skin and finding the artery. Accordingly I proceeded by ambulance to Llandudno hospital. Here I spent the night quite comfortably except for my rapidly increasing hunger. It is apparently not done to give patients anything to eat after nine o'clock. At four a.m. the night nurse was finally constrained upon to provide some bread and butter. I have since then been very well treated. I have here no money, no spectacles, and a most peculiar outfit of clothes, but I shall get to Helyg and collect my belongings in any case. This hospital is a marvellous modern place, reminiscent of Aldous Huxley's "Park Lane Hospital for the Dying" in *Brave New World* [1932]. Wireless laid on for every bed. Do not worry about me, I shall be out and about by the time you can communicate with me.

In Huxley's vision of the future, the fear of dying has been abolished by universal use of euphoric drugs, and boredom has been abolished by ubiquitous television.

During the time that I spent at the climbing hut in Wales, a week at the beginning of December 1941, the big world changed. Japan attacked Pearl Harbor and declared war on the United States and Britain. Hitler declared war on the United States. Suddenly we had the Americans as allies. The victorious end of the war changed from a distant hope to a certainty. We heard nothing of these events at Helyg. I learned that the world had changed only after I was back with my family. There is not a word in the letters about the changes. Life at Cambridge continued as before.

JANUARY 24, 1942

I went along to Besicovitch to see him and found him at the billiard table and very friendly. Having found when the lectures were, I made so bold as to ask him about Russian teaching, and he appeared to be very interested and began talking volubly in Russian. Remarking desperately "*medlenno, pozhaluista,*" I stemmed the flow and managed to collect my wits sufficiently to conduct a conversation of sorts for a few minutes, in which it was decided that I should go to Prince Obolensky who is a Russian research student in Trinity, and he would introduce me to Miss Hill the boss, who only comes down from London twice a week. He then offered to write out a line of introduction for Prince Obolensky and produced a page of illegible writing for me to present to him. As the Prince is not yet in residence, but is returning over the weekend, that is as far as the matter has yet progressed. On Wednesday I went to Littlewood for the first time and liked him very much. He is talking about the modern view of functions of a complex variable, and though I know most of it vaguely, he makes it far more concrete than the books do. It is very pleasant to find such a lot of people who are glad to see me. Even Hardy asked me whether it was a bomb or a motor accident, and Besicovitch told me a grim story about how when he was young in Russia a friend in jest plunged a three-bladed knife into his back, after which he lingered on death's door for a fortnight.

WEDNESDAY

The noble Prince Obolensky continues to inhabit the Absolute Elsewhere, and so I have not made contact with him yet, but I have had an even greater honour, for Besicovitch invited me in on Monday night to have my accent superintended, and he started off by making me repeat after him line by line a little poem of eight lines until I knew it by heart, and explained where the mistakes were. I continued this for about forty minutes, and I have no doubt I have benefited. He says that the general effect is not bad, probably owing to the Linguaphone records at Winchester. The difficulty is always in pronouncing the consonants gently enough, for to a Russian an English v is very nearly an f and a g nearly a k and a th nearly a t. And one thing which I was very nearly beaten by was the difference between hard sh and soft sh. However it was very amusing, and it is a lovely little poem. After that was over we conversed a little in Russian, and finally he found it easier to talk in English and talked very interestingly about various things until I departed. He confessed that he was finding it difficult to keep to his new year resolution of only six hours billiards a week, and that the other day Hardy made a break of eighty-two; but of course I was most interested in his dealings with the Bolsheviks. He said that after the revolution he organized a high school at Perm and for three years successfully resisted the Bolshevist doctrines of education by telling the teachers to teach exactly in the way they thought best. The official doctrine of education is that everything must be taught in relation to labour, and that it is the job of the teacher to discover how this is to be done; so if it can't be done, the teacher has to pretend it can. However he said the Communists in practice usually had a great respect for academic learning, and they came to his school and worked hard.

After that he went with four professorships at a total stipend of 150 pounds a year to Leningrad and taught for two or three years. The Communists there being more regimented, they quarrelled with him and ousted him one by one from three of his four positions. He said that as a mathematician he could always have got support from the Moscow government, but at this point he left Russia. The Communists have since then gradually come round to sensible methods of teaching. He said that Bolshevism alone has saved Russia now and has been the effective cure for the country's diseases. But I imagine that he left because he put mathematics first, both ideally and from personal preference. I did not press him to ask whether the story of his escape by swimming the Volga was true.

Besicovitch was lucky to leave Russia before the great purges of the 1930s, when a large fraction of educated people were executed or sent to concentration camps. Many of those who survived the camps were released when Russia was invaded in 1941. His view of Russia left out most of the worst horrors.

Thursday

Prince Obolensky, I discovered last Sunday, after standing about for an hour or more in bitter snow and wind, is working in London and only comes to Cambridge on Sundays once a fortnight. Then he takes his meals with friends and comes to his rooms just to pick up his belongings. So that was why I had to stand about; I finally extracted that much information from his bedmaker, but it all seems pretty useless. The famous footballer Prince Obolensky who died recently was the cousin of this one, but as they derive their title in the direct line from Rurik the first Lord of Russia around 900 A.D. before tsars were invented, it is not surprising that there should be more than one.

February 10, 1942

I received a cake from Aunt Margaret last week. I managed to invite a man to eat it with me on Sunday, and I chose a refugee by the name of Chrysel [*Correctly spelled, his name was Georg Kreisel. He played a dramatic role in my story fifteen years later.*] who has come up this term and is a pursuer of "higher thought" in matters mathematical. The tea party was not entirely a failure, and the guest was very appreciative. He talked incessantly about the nature of knowledge. He goes to some classes held by Professor Wittgenstein of which I was entirely uninformed, but which sound most amusing, though Chrysel like a good German is incapable of taking anything except with deadly earnest. How these Germans take things seriously. Chrysel remarked casually that he had read fifteen philosophical books by Bertrand Russell alone since he came to England two and a half years ago, and of course he has read one at least of Kant and Hegel, Descartes and Leibnitz in the originals, three or four of Whitehead, three of Moore, and any number of others. He was at Dulwich School and apparently picked up his philosophical bent there from his housemaster. He is taking mathematics as his main subject and knows a good deal about that too. He lived in Vienna, but I do not know what else he has done. He is a logical positivist. He makes his creed "Everybody else is wrong" rather than "I am right," like Socrates, and is equally annoying if you are not in a good temper, as Socrates's victims usually were not. Fortunately I was in a good temper on Sunday and enjoyed it very much. I shall go to tea with him one of these days.

The other thing I have achieved since last week is to make my first contact with Elizabeth Hill. It is amazing that it has taken a month's effort to do it, and I am not through until next Tuesday. First I tried Obolensky, and it took two weeks' search to discover what had happened to him; then I asked Besicovitch how to find Miss Hill, and he said, "See her at her lecture room in Mill Lane." On Monday I was going down to Mill Lane in a vague hope that I might discover where her rooms were, and I happened to go into a bookshop on the way, and what should I hear but a young lady ordering some Russian books, and she said, "But why haven't you got the things that Miss Hill ordered three weeks ago?" So I took the heaven-sent opportunity and said, "Can you tell me Miss Hill's name and number?" and she told me her address, which wasn't in Mill Lane at all. It is marvellous how inaccessible people manage to be.

WEDNESDAY

Tuesday being the great day when I was to meet Miss Hill, I repaired to Mill Lane in good time to see her at nine and found a large class of people, and when Miss Hill came in, she singled me out and said she was glad to see me. She must have been told about me by Besicovitch. The class was fairly elementary, but it is definitely worth going to for vocabulary and points of idiom if nothing else. She began with half an hour's communal translation from English to Russian, and then half an hour back again. She is a good teacher, and I picked up quite a few new ideas during that hour. I found Henry Padfield, a man whom I knew vaguely at Winchester. They expect me to take part in their performance of the Kossovo Ballads at the end of the term. It is being recorded for a broadcast on Kossovo Day sometime in June. She says that I shall probably have to appear in a crowd of warriors or peasants or something, but I have not heard any more as yet.

FRIDAY

I am still going to Mill Lane at nine in the morning on Tuesdays and Thursdays and like it very much; I am also going definitely to appear as one of a crowd in the Slavonic Society's Yugoslavian binge. That was not what I intended, but it will be amusing. The performance is next Wednesday so that they do not require very much from me in the way of going to rehearsals; as far as I know there is one on Sunday evening and probably only one after that. So it will not be at all burdensome. I went to Besicovitch last Wednesday. I found there a young man to whom Besicovitch had given similar instruction a year or two ago; it appears that Besicovitch is in the habit of improving the Russian pronunciation of anyone he can get hold of. At any rate this young man had come back to see Besicovitch, and he recited the same verses that I had begun reciting to him. So we all three sat and conversed in Russian, and I did very well; it is so much easier with three people, as you can let the other two carry on while you get your next remark straight. He said that he was in the army in an officers' educational course and was hoping to get to Russia. Although he speaks the language quite fluently, he said that it was very unlikely that they would have him in that capacity. They are, however, offering special scholarships to any students who are prepared to learn Japanese or Persian.

Last Wednesday, a week later, Besicovitch asked me to come again and discuss one or two mathematical problems. I arrived at eight-thirty in my usual clothes and found a regular social gathering with subfusc suits and long dresses. I was just preparing to run away and hope nobody had seen me, when Besicovitch welcomed me, and we sat down on a sofa and talked mathematics while the others played bridge. Besicovitch has the most perfect manners of anybody I have ever met, and made me feel more or less comfortable. He always introduces the guests, and escorts them to the door with a low bow, and generally behaves like a landed aristocrat. Mrs. Besicovitch was there, also a Miss Thompson and four more people who played bridge with each other, and whom I did not see much of except for handing round toast and tea at ten o'clock. Besicovitch was as amusing as usual in spite of the uproar. He tried to introduce Miss Thompson into the circle, but she is not a good linguist; she has been learning Russian from Mrs. Besicovitch, she said, for years, but she could not understand anything Besicovitch, still less I, said; so we finally relapsed into English. I shall be very interested to know more about Besicovitch, as I hope I shall in time. When he asked Miss Thompson whether she, having toured Lapland, could speak Lappish, I asked him whether he, having lived at Perm, could speak Tartar. He said he was surprised that I should have thought of that, as his native language was actually Crim Tartar. His family lived in Crim Tartary and spoke Crim Tartar among themselves, including his elder brothers and sisters, but when he was born, they decided only to speak Russian to him so as not to spoil his chances of speaking Russian like a native. I never thought when I read *The Rose and the Ring* that such a place as Crim Tartary existed, or that I should ever meet a native from it. But what sort of people his family were he did not say. As far as I remember, in the book the sympathy of the reader was not on the side of Crim Tartary.

Thackeray's The Rose and the Ring: A Fire-side Pantomime for Great and Small Children *(1854), was one of the favorite books of my childhood. Written under the pseudonym Mr. M. A. Titmarsh, it is a spoof of a romantic fairy story, involving the warring states Paphlagonia and Crim Tartary.*

TUESDAY

The Kossovo play was a success, beyond all my expectations. I was told to go to a rehearsal last Sunday week at eight-fifteen p.m. and after waiting in the dark for half an hour discovered that that rehearsal was really at two-thirty in the afternoon. So I came for the dress rehearsal on Monday without any previous experience; and that dress rehearsal lasted from four till eight, during which time three out of ten scenes were rehearsed; it was about the most incompetently managed rehearsal I have ever attended. However, on Wednesday night when the King and Queen of Yugoslavia and the entire staff of the Yugoslav forces were situated in the front row, and the BBC had microphones and cameras all over the place, things were fortunately very different. The main difference was caused by a handsome and clever young man who took the part of the Bard, who was not performing on Monday. The play being a dramatised version of ballads, there is a great deal of monologue by the Bard while the other people sit in silence. Thus it was a great asset to find that this young man had a marvellous voice, which made the rather indifferent versification sound highly poetic. This man was called Dimitri by everybody, and it was not till afterwards that I discovered he was the long-lost Prince Obolensky. Apart from him the chief part was Tsar Lazar, played by Henry Padfield whom you may remember from Winchester. Padfield was considerably brighter, since he went with the Home Guard on their twenty-eight-mile night march on Saturday night, and had recovered by Wednesday. The performance on Wednesday was very effective and was broadcast to America and to Yugoslavia during the week, and the queen of Yugoslavia was in tears all the time. The BBC sent Miss Hill a letter of congratulation which she read out proudly this morning.

On Thursday there were no microphones and cameras, but otherwise it was the same, and again was successful; at the end the actors were invited by the king to a party in the Arts Theatre, to go on till one o'clock. The ballads of Kossovo are very interesting and well worth listening to; I do not imagine anything is known about their authorship, but it is a beautiful story. Before the battle St. Mary sent a falcon from Jerusalem to the king, with the message that he must choose either to win a great victory or to reign in a heavenly kingdom, in which case he was to draw up his whole army to a solemn Mass on the field of Kossovo, and then they would know certainly that the whole army would perish.

King Peter was in exile in London, still recognized by the British government as the lawful sovereign of Yugoslavia. His agent General Mihailovich, in Yugoslavia, was supposed to be organizing resistance to the German occupation. Meanwhile the Communists led by Marshal Tito were organizing a more effective resistance. The British government later transferred recognition from the king to Tito.

I am now comparatively idle; this morning Dirac gave his last lecture, so that will be the last time I shall have four lectures in a morning. I have found that listening to all these lectures is just about as much as I can do; Dirac reached a climax of difficulty in his last lecture. Littlewood has also finished, so I have only one more of Miss Hill, Hardy, Newman, and two of Besicovitch.

Monday

Besicovitch is without doubt a complete dear. On Wednesday he invited me to come with him after lunch for a walk to speak Russian, and we talked about some problems he had given me a few weeks ago. I told him a few incipient ideas I had had on the subject, and he was so encouraging that I spent the next two days in a fervour of activity, and actually proved two quite interesting results of which one may be new. This is the way I have always wanted to spend my days, but it is seldom that it really happens. On Friday the weather was lovely and all Besicovitch's supervising was over, and he took me for another and longer walk round the country, during which he discoursed largely on Russia; and Friday, Saturday, and Sunday I spent an idyllic existence, doing research in the mornings and evenings and talking Russian in the afternoons. Indeed I have been talking so little English this week that I shall soon be forgetting it.

It appears that I am not the only person on whom Besicovitch lavishes his goodwill, for there was a most unexpected incident on Saturday when we came in to Neville's Court on our return; a small figure at the other end of the court outside Besicovitch's rooms rushed down the court straight at him, and he shouting "My darling" embraced her right off her feet. She was about twelve years old, but as they went to his rooms then I did not discover who she was. On Sunday, however, as we were returning through the backs we met another young girl, rather smaller this time, plus a father. The three of them talked a very peculiar conversation, the girl to her father in Czech, the girl to Besicovitch in English, and Besicovitch to the father in Russian. Besicovitch was if possible even more delightful with this young friend than with the other, and they chased each other round the lawns to the great delight of the many passersby. It appeared that both of the girls were Czech refugees, but they were not related; the second one was called Ladochka, lived at Brno, and her father escaped sometime after Hitler arrived by leading a trade mission to Rumania and taking his family with him. They both spoke perfect English, though the father did not. It is certainly no wonder that people are so fond of Besicovitch.

I have had a Russian game of billiards with Besicovitch. In the billiards I did very well on the whole, making 109 while he made 163; he is fairly good. Hardy the other day made a new record break of 119 and was enormously pleased with himself.

MAY 19, 1942

In my second year as an undergraduate at Cambridge, I registered for national service. I was in the Home Guard, which was supposed to help the army defeat the Germans in case of an invasion. In 1940 the threat of an invasion had been real, but in 1942 the Germans were heavily engaged in Russia, and nobody took the threat of an invasion seriously. It was lucky that we never had to do any real fighting. The Home Guard required me to take part in occasional night exercises. The invasion became more and more unlikely as the years went by, but the exercises continued until the end of the war. They helped to sustain the wartime spirit that made England a friendlier country during the war years.

I went on the exercise on Saturday after all. We sat in the parade ground till ten p.m., when my section set out for Grantchester on bicycles. The idea was to stop the Welsh Fusileers from capturing Cambridge, so we were stationed near a bridge at Grantchester; two men to demolish the bridge and the rest to give protection to them. We took up a very halfhearted defensive position on the roadside, lying along the hedge, and waited for orders to demolish the bridge, or the arrival of the enemy. We had a beautiful notice saying Bridge Demolished and a small piece of explosive to make a bang. Of course, as always when I am on an exercise, the enemy never came anywhere near Grantchester, though they overran about a third of Cambridge. We stayed under that hedge from eleven p.m. till eleven a.m. It was impossible to take enough clothes to be able to keep warm, though the weather was quite fine. Last night I spent at the hut fire-watching. I was out from twelve till two-thirty but slept pretty well the rest of the time. Fortunately the weather was again good. But this military activity does take up a great deal of energy when they have night operations.

FRIDAY

The other evening I played a game of billiards with Besicovitch and had the honour to be directed in the strokes I should play by Hardy, who is a first-class player. He sat in an armchair and criticised both of us. In the end he arose and said that he would enter the game on the seventh break from now and would play one break; if the score he made was of the form $4n+1$, it would go to me, if of the form $4n+3$, it would go to Besicovitch, and if it was even, it would be null and void. He urged us to play so as to leave either a good or bad position as suited our scores at the moment. When the moment came, he scored zero, and so no ill-feeling was caused. In the end I won the game.

NOVEMBER 10, 1942

Since Thursday I have spent most of my time working on the theory of partitions. I discovered, by experiment, a striking result in the theory; and I have been trying to prove it, without success, and when I got tired of trying to prove it, I started verifying the result for various special cases, a process which can go on ad infinitum; so I have been rather busy. The partitions of 4 are five in number, namely 4, $3+1$, $2+2$, $2+1+1$, $1+1+1+1$. Ramanujan discovered that the number of partitions of any of the numbers 4, 9, 14, 19, 24, 29, is always divisible by 5. If in each of the partitions of 4 you take the number of parts and subtract it from the largest part, you get in the five cases $4-1=3$,

3−2=1, 2−2=0, 2−3=−1, 1−4=−3. You have the five numbers 3, 1, 0, −1, −3, and one of them (0) differs from 0 by a multiple of 5, one of them (1) differs from 1 by a multiple of 5, one of them (-3) differs from 2 by a multiple of 5, one (3) differs from 3 by a multiple of five, and one (-1) differs from 4 by a multiple of 5. What I have discovered is the following: If N is any one of the numbers 4, 9, 14, 19,—and you write down all the partitions of N, you can treat each of them as I did before, i.e., you can subtract the number of parts from the largest part. You get a set of numbers, which you can arrange in five groups according as they differ from 0, 1, 2, 3, or 4 by a multiple of 5. In the case N=4 there is just one number in each group. I find that there is always an equal number in each of the five groups; for instance, when N=9 there are just six in each group. But I cannot prove it in general. You will see that this result explains why the total number of partitions is always divisible by 5, because the total number of partitions is got by adding the membership of all the five groups.

The partitions of 5 are seven in number, namely 5, 4+1, 3+2, 3+1+1, 2+2+1, 2+1+1+1, 1+1+1+1+1. Ramanujan proved that the number of partitions of any of the numbers 5, 12, 19, 26, 31, 38 is always divisible by 7. My result then carries over word for word to this new case; you get seven equal groups instead of five, when N is one of the numbers 5, 12, 19, 26, 31, I have verified this for N=19 which comes under both cases. You get 490 partitions, and from these you get 490 numbers by subtracting the number of parts in each from the greatest part. Then if you divide them by the scheme 0, 1, 2, 3, 4 you get five equal groups of 98 numbers. If you divide them by the scheme 0, 1, 2, 3, 4, 5, 6 you get seven equal groups of 70.

I published my discovery about partitions in the student magazine Eureka *which was the journal of the Archimedean Society (1944). The conjecture which I had verified numerically was finally proved analytically in 1953 by my friends Oliver Atkin and Peter Swinnerton-Dyer.*

NOVEMBER 17, 1942

I made a closer acquaintance with an undergraduate called [Fritz] Ursell whom I like very much. We had a session on Sunday night, he providing coffee and I cake. He is a refugee from Düsseldorf, and talked a lot about the refugees. He confirmed me in the opinion that the policy of the English government and trade unions toward refugees is both uncharitable and foolish. In particular I was impressed by the fate of the majority of his family who stayed in Germany; they are now all in a ghetto in Lublin. There was only the choice between clearing out altogether and complete defeatism. It was not the weaker elements that became refugees, but all the people who had sufficient energy and moral courage to wish to preserve their standards, and to carry on their lives, at the cost of heavy and immediate sacrifices. However, I will not preach any further. Ursell's father is a quite eminent doctor, and he is not allowed to practise; though in fact there is and always will be enough work for any quantity of doctors.

It is difficult to remember how much we knew or cared about the deliberate killing of Jews by the German government. We certainly knew that Jews were persecuted and dismissed from their jobs. Fritz Ursell was one of several Jewish friends who had escaped to England and told me how badly their relatives were treated both by the Germans and by the British. The full extent of the horrors in the concentration camps became known only as the camps were overrun in the last year of the war. Hitler's decision to implement the Final Solution and massacre all the Jews was kept secret in Germany and was not known to us in England until the end of the war. Both in Britain and in the United States, the governments had a deliberate policy of suppressing information about the killing of Jews, knowing that in both countries the quickest way to lose public support for the war would be to give the impression that we were fighting the war to save the Jews. Neither Churchill nor Roosevelt was fighting the war to save the Jews.

NOVEMBER 30, 1942

On Saturday I went to tea with Besicovitch and enjoyed it more than any tea party I have attended for a long time. There were present two young Diracs, by name Gabriel and Judith. Gabriel is seventeen and a first-year undergraduate from St. John's, Judith fifteen and at a school in Cambridge. They are Hungarians by upbringing and knew a lot about Central Europe; also they used to row on the Danube with von Neumann the great topologist, an almost legendary figure now in America. They came to England about four years ago, but speak the language very well. Gabriel is an ardent member of the Communist Party and kept the ball rolling from the start. Besicovitch himself is not a Communist but is the most persuasive arguer for it that I have ever struck, as he never makes wild statements or resorts to political clichés. The young Dirac is reading mathematics but is at present more interested in politics. I sometimes think I ought to become more political; it is a very fruitless activity but it does keep one from getting bored. Ursell the refugee was telling me that the real crisis of reaction in Germany was in 1921–22 when all the competent Communist leaders were murdered. He is however of the opinion that Russia will keep them from doing it again this time. I hope so, as I am sure England would not raise any objection. On Saturday I played a game of billiards with Hardy. He is very good. He plays incredibly fast; made a break of 53 in about three or four minutes. The score was about 240–140, and I had also a heavy handicap.

The young Diracs were stepchildren of the physicist Dirac, children of the physicist's wife by a previous marriage. They took Dirac's name when he adopted them.

2

WAR AND PEACE

SOMETIMES A GERMAN BOMBER, *having lost its way, would fly over Cambridge and drop a couple of bombs. One of these bombs fell on the student union just across the street from my bedroom in Trinity College. Since I belonged to our college fire service, I was ready to spring to action. But the college authorities told us that our job was to protect the historic buildings of the college. We were not allowed to cross the street. So we stood idly watching while the union building burned down.*

Notable characters in this chapter are:

Charles Percy Snow, then thirty-eight years old, was technical director of the Ministry of Labour, afterwards famous as a writer of novels and social criticism. He was then a bureaucrat, responsible for assigning technically trained students to wartime jobs. He made the decision to assign me to the Operational Research Section of the Royal Air Force Bomber Command, starting in July 1943, when my two-year deferment at Trinity College ended.

John Maynard Keynes, then fifty-nine years old, was effectively running the British economy during World War II. He showed his mastery of economics at three levels, first by becoming wealthy himself, second by making King's College the wealthiest college in Cambridge, and third by keeping the British economy afloat through the stormy years of war. He died in 1946, worn out by his efforts to manage a fourth even bigger problem, the reconstruction of international economic institutions after the war.

Henry Moseley died at the age of twenty-seven in 1915, before the scientists of my generation were born. By his death, he saved our lives. He was a brilliant young physicist who had been a student of Ernest Rutherford in Manchester. Rutherford was then leading the world in the science that became nuclear physics. He had done the crucial experiment that proved that every atom has a nucleus. In 1913 Moseley made his great discovery, a simple law relating the wave-length of X-rays emitted by an atom to the number of electrons in the atom. When World War I broke out in 1914, Moseley volunteered and enlisted in the army, went to fight in the disastrous Gallipoli campaign in Turkey, and was killed. After that war ended, the death of Moseley was recognized as a tragic waste of a rare talent, and the British government decided that in future wars promising young scientists should be kept alive. That is why I was put into a safe job at Bomber Command headquarters while other boys of my age flew in the bombers and were killed. Two of my close friends were killed in the last year of the war while I was at Bomber Command. The gross unfairness of the system left me with a permanently bad conscience for having survived.

January 24, 1943

Twice I played billiards with Besicovitch, and once won against him (though with a handicap). Each time Hardy came in and took over from Besicovitch; he is teaching me quite a lot about the game; the only trouble is that after an hour of Besicovitch and half an hour of Hardy, I am so exhausted that I miss the ball completely. However I enjoy it very much, and Hardy seems to too, as he seldom finds anybody to play with. Besicovitch refuses ever to play against him, and his only real equal is a don called de Navarro, a lecturer in modern languages, who is sometimes there when I go to see Besicovitch.

The Germans bombed London heavily in 1940–41 and less heavily with V-1 cruise missiles and V-2 rockets in 1944–45. During the years 1942–43 there was very little bombing. Occasionally a few bombers would come over and drop a few bombs, probably to reassure the German home population who were suffering from the growing British and American bombing of Germany. One of the token raids on London happened in January 1943.

I wish I had been in London for the air raid; there is nothing that makes me so happy as a display of fireworks. It seems they have given up the idea of a Baedeker raid on Cambridge, which is a pity. I bought yesterday a set of *War and Peace* in Russian, the first I have seen in England, published in Moscow in 1941. Rather a wartime production I am afraid, but still readable and with quite large print. I shall settle down to it one day, not in the near future. It is 1,927 pages long. It costs eighteen roubles in Russia, and a pound in England, the difference being several hundred percent; I do not know where the money goes. It says they published 100,000 copies of the edition in 1941–42, evidently as in England there was a great demand for it. I read the concluding paragraphs, which point the moral of the whole work, that human affairs can only be understood by a belief in complete dependence on Providence. They have not, I am glad to say, been interfered with.

I still have on my bookshelf the wartime Russian edition of War and Peace, *with wartime paper now brown and brittle. The final paragraph makes a comparison between the ancient astronomical belief in a stationary Earth and the modern belief in human free will. Tolstoy says that both beliefs are illusions. In astronomy, the earth moves, and in human affairs, we are inescapably dependent on providence. The Soviet government in 1941 allowed this radically non-Marxist state-*

ment to be published. The Russian people in those days took more comfort from Tolstoy than from Marx.

JANUARY 30, 1943

The other day we had an interesting talk by Littlewood to the Adams Society, on study and research in mathematics, talking about the various techniques of working. He is a marvellous lecturer, so good that he only has to stand in front of the blackboard and say nothing for the audience to burst into roars of laughter. He has made a scientific survey of the effects on himself and his pupils of various habits of work; and he has by now brought the art of research to a high degree of perfection. His own habits are illuminating; he always goes down to Cornwall to the coast for any intensive period of research, and allows no extraneous intellectual strain to disturb him. When he is there, he keeps rigidly to a maximum of 5½ hours a day, and 5½ days a week. The rest of the time is spent rock-climbing, walking, and reading detective stories. Also, whenever he has a bright idea, he immediately stops work for the rest of the day; very frequently this causes another bright idea to occur at the same time on the next day. Another rule is, at least three weeks of continuous complete holiday a year, preferably spent in difficult mountaineering. Also, whenever you start a cold, you should take a complete holiday till it is over. He further discussed the virtues of alcohol and tobacco. He said that to him they both decreased productivity by about one half. At these words he produced a gigantic cigar, lit it, and said, Ah, but it is one of my rules never to work after dinner.

Littlewood was a great joker. On another occasion when he was giving a mathematical lecture, he arrived with his beloved cat resting on his shoulder. All through the lecture, the cat stayed on his shoulder, sometimes sleeping and sometimes walking from one side of his head to the other. The students had a hard time trying to concentrate on the substance of the lecture.

The fact that Littlewood changed his schedule for my benefit is surprising. I suppose that the famous professors were starved for students and valued our listening as much as we valued their talking. My friend James Lighthill and I came from Winchester to take the Trinity College entrance scholarship examination in 1939 when we were fifteen years old and both won scholarships. The college decided that we were too young to be students, so our entrance was delayed to 1941. The professors no doubt heard about this and were expecting something unusual when we arrived.

I am also surprised that I found Arthur Eddington's course the least worth hearing. Eddington was the astronomer who observed the displacement of star images by the sun during the total eclipse of 1919 and proved that Einstein's theory of general relativity was correct. I expressed my low opinion of Eddington's lectures at the beginning. In retrospect I remember the course as one of the best. By the time it ended, I had become a personal friend of Eddington.

I am now going to nine courses of lectures; this is due to the fact that Littlewood refused to go on lecturing unless I should come, and even took the drastic step of changing his time to nine a.m. so that I should have no excuse for not coming. Accordingly I now go at nine a.m. and am very glad because the lectures are the best of my nine courses; Hardy this year is not as good as he was last year, as the stuff he is doing is less monumental. All the courses are well worth hearing. Eddington's relativity perhaps the least so, as I know the stuff already; but as there are only three of us, we get a good deal of opportunity for asking questions and getting valuable tidbits of one kind and another. Jeffreys on aerofoils is much less deadly than I expected. This morning he was discussing an ingenious method of calculating the effectiveness of various wing-shapes of an aeroplane; usually one calculates the pressure of the air on the aeroplane, but in this method you ignore the aeroplane and get the same answer by calculating the pressure of the air on the ground as the aeroplane passes over.

Enough of mathematics! Last night we all went to a talk by Dr.

Snow of the Ministry of Labour about our careers. The government this year will fix people up as soon as possible during the next four months, so that they will go to work immediately after their examinations are over, except in case of failure. So I should know my fate sometime before next June. The Services are establishing what they call operational rather than material research, i.e., research in the use of weapons on the basis of collecting and analysing the experience of actual fighting, and giving it statistical treatment. For this they want only the best people, as the job involves not only technical knowledge but the ability to extract information and give advice to generals. The demand for radiolocation is small now, except for U-boat outfits. He said the government had been caught out by the rise in the production of U-boats since the invasion of Russia, and that everything connected with U-boat-hunting had first priority. The Germans using the methods of operational research were just now discovering how to use large numbers of U-boats to the greatest effect. The demand for operational research workers on our side would rise sharply when England begins serious fighting. As far as our army was concerned, the Middle East campaign counts for very little.

This afternoon I went to see the Joint Recruiting Board. Dr. Snow did most of the talking, and it only took about five minutes. The board after asking perfunctory questions asked me what I wanted to do; I said I did not mind but would prefer something active. They said that they would give me the best job at their disposal in operational research. They could not promise me anything as it depended on the services whom they finally chose. They said it would probably be a civilian job at first, but if I was any good at it, I should get a commission. Also I should likely go abroad soon. They were pleased at my learning Russian but agreed that it would not probably be any use.

I asked Littlewood at the meeting on Tuesday, whether he ever had any trouble in his work, to find problems to research upon. This has always seemed to me the most difficult part of the whole business. He made a noncommittal answer. Then yesterday after the lecture he came up to me and said I must not worry about finding problems for research, that research never mixes well with learning. He said that once learning was finished and you get out on top of mathematics, there is a complete abundance of problems; and he had never had any difficulty, nor his pupils either. I was very pleased to know this, and still more pleased that he should have spontaneously talked to me about it with such earnestness.

One other thing Dr. Snow said, that all war jobs are absolutely foul, and that no one should go into it with rosy and romantic ideas. So I am not expecting to find my position either congenial or important; but it is difficult to dispel rosy illusions until they are knocked out of you.

FEBRUARY 7, 1943

I have just had an interesting talk with an undergraduate called Oskar Hahn who plays the flute in the orchestra. It turns out that he is a nephew of Kurt Hahn and spent five years at Gordonstoun. He is extremely enthusiastic about that school, and I think it does sound about all that can be desired. In particular I like the idea of only having two afternoons of games a week and otherwise doing constructive activities. Whether I should have enjoyed it, I do not know; it is decidedly nonintellectual. It is designed for the ordinary people, not for the intellectuals; instead of giving scholarships by competitive examination, they give them to the children of the local fishermen. Hahn said that they get extraordinarily good results from almost everybody. He is a cripple, having one leg completely paralysed from infantile paralysis (polio) at the age of ten. But that did not handicap him much. For sailing, and most crafts, all you need is a good pair of arms. I shall try to see more of him now that I have discovered his name.

Kurt Hahn was a German Jew who founded a boarding school in Germany with the name Salem, intended to educate an elite group of children in a counter-

culture opposed to the prevailing academic culture. The school flourished under the Weimar Republic, attracting children from aristocratic German families as well as Jews. In 1933 Kurt had to leave Germany and quickly founded a similar school at Gordonstoun in Scotland. Gordonstoun was equally successful. Several members of the British royal family were educated there. After the war, Kurt returned to Germany and revived the Salem school. Gordonstoun and Salem are both still flourishing. I never met Kurt, but I heard many stories about him from Oskar Hahn. Oskar said that Onkel Kurt was a pompous ass but was also a wonderful teacher and organizer.

Last Sunday we had a very interesting talk on Newton by Maynard Keynes [1946]. He is a great expert on Newton and has gone through a vast amount of unpublished work of Newton's on every subject under the sun. He spoke extremely well, and his conclusions were amazing. The common picture of Newton, as the great analytical mind sifting the evidence and being satisfied with nothing short of mathematical certainty, is completely wrong. He was essentially a magician rather than a scientist, and all his work was inspired by his studies of the old alchemists, metaphysicians, and apocalyptics. It was known already, said Keynes, that Newton was interested in alchemy and the Book of Daniel; but his papers show that he considered all his activities as in the nature of solving ancient riddles rather than discovering new facts. With any luck, Keynes will write a book about it after the war. Another remarkable feature of the talk was Newton's perseverance: he used to work incessantly from early morning till late at night, month after month and year after year, until he had a breakdown at the age of fifty. After that he gave up work entirely and became a member of society. It also appeared in the course of the talk that the man who allowed Newton's papers to be sold in public auction, so that many of them are now lost, was none other than our old enemy Lord Lymington. I think that is a perfect example of Nemesis, that a man of his character should be branded by Providence with such a supremely suitable and permanent notoriety.

Lord Lymington was a notorious admirer of Hitler who owned a large piece of land near to Winchester. He would probably have become our Gauleiter if England had been successfully invaded. I had got to know him while working to bring in the harvest on his land. During the war, most of the young farm workers were away, and high school kids spent several weeks of their summer vacation loading wheat and oats with pitchforks onto trucks. Lord Lymington had us as a captive audience and told us about the kids he had seen in Germany working in the Kraft durch Freude (Strength Through Joy) movement organized by Hitler. He told us that the German kids worked better because they had an accordionneuse, *a young girl with an accordion who kept them working to her rhythm. He promised to send us an* accordionneuse *too. Fortunately, the* accordionneuse *never showed up. We would probably have given her a bad time.*

The summers of 1940 and 1941, when I worked for Lord Lymington, were both wet, and we worked much of the time in pouring rain, collecting sheaves of wheat and oats that were already ruined, with green sprouts growing out of the rain-soaked grain. A large fraction of the harvest was lost. Farmers in England were accustomed to such losses and did not know that they were easily preventable. The farms were kept alive through the war by government subsidies. After the war, the farmers found the cure for rainy summers. The cure was drying sheds, developed by tobacco farmers in Virginia. The sheds are cheap to build and need to be heated for only a few days to dry a rain-soaked harvest. Drying sheds transformed farming in England from a losing to a profitable business. So far as I know, although we were clever high school kids, none of us ever thought of this simple solution to the farmers' problem.

SUNDAY

I spent last evening talking to Georg Kreisel, the Viennese refugee whom I may have mentioned before as being one of the ablest people here. It is remarkable, but I am finding the refugees on the whole the nicest people. In particular, Kreisel is an outstandingly solid character. It makes one feel very small to talk to a person who has been in Dachau concentration camp; Kreisel was there for a fortnight when he was fifteen and was spared none of the ghastly details.

His parents are at present living in our concentration camp at Mauritius, and likely to stay there for the duration. He lives in England almost entirely on his scholarships. Fortunately Trinity College does not as a rule discriminate against foreigners in the election of fellows, and I think he has a good chance of a fellowship which would assure his future. He has no bitterness against anybody; I, on the other hand, get all hot with anger when I think of the incredible lack of human sympathy with which the English allowed these persecutions to be hushed up, and now extend so grudging a welcome to the victims. Please excuse sermon.

My mention of the British concentration camp in Mauritius does not imply that it was equivalent to Dachau. Mauritius is an island in the Indian Ocean that was then a convenient dumping ground for aliens who had been denied entrance to Britain. Aliens were free to live and travel as they pleased on the island but were not allowed to leave it. It was technically correct to call it a concentration camp. The aliens there were effectively in prison, even if they could move around. Mauritius then was similar to Australia 150 years earlier, a humane alternative to hanging.

My protest against the treatment of refugees was directed against my father, who had been president of the Incorporated Society of Musicians, the musicians' trade union, before the war. As president, he had acted vigorously to keep refugee musicians out of Britain and to prevent them from taking jobs away from British musicians. One case that I remember vividly was that of Rudolf Bing, a gifted conductor who had taken refuge in London. My father succeeded in blocking him from finding a permanent job in England. He was forced to emigrate to New York, where he enjoyed a brilliant career as conductor of the Metropolitan Opera. My father did not regret his action. My father said, "The Americans can afford Rudolf Bing. We can't."

MONDAY

The lectures of Jeffreys are great fun. The audience is composed of me only, and Jeffreys has stopped doing aerofoils and is lecturing on all sorts of recondite problems that he has solved during his life-time. One of these problems is the cooking of porridge in a shallow saucepan. According to the classical mechanics of viscous fluids, the convection currents should always prevent the porridge from over-heating at the bottom. The new theory due to Jeffreys, which is very ingenious, shows that stirring will in general be necessary to avoid burning, if the viscosity of the porridge exceeds 6,000. I hope this may alleviate your domestic problems.

Harold Jeffreys was a geophysicist with a wide range of interests, including the interior of the earth and the dynamics of the atmosphere and oceans. A large part of his course was concerned with ocean tides and their effect on the orbit of the moon. I was always amazed that he took so much trouble to explain these wonders of nature to a single undergraduate.

This afternoon I spent at the cinema. The programme was *Citizen Kane* followed by *The Ghost Train*. *Citizen Kane* is definitely worth going to see, if you have not already done so. It is the story of an American millionaire, done with incredible artistry and sympathy. You may remember my being enthusiastic last year about a French film, *L'Atalante*; a few weeks ago there was an article in the Sunday *Times* devoted to it, which praised it ecstatically. *Citizen Kane* has some of the same virtues, though it suffers more from the vice of ambitiousness. It was made in England in 1941, which is a cheering thought. The other night we read *Heartbreak House* aloud; I had the part of the old captain, and enjoyed it immensely.

Heartbreak House is a play by George Bernard Shaw about England in World War I, with Captain Shotover as the most memorable character. It had a powerful resonance for the generation growing up in England in World War II.

I was agreeably surprised on Thursday to receive a large envelope stamped Princeton, February 11, 1943, and inside it, lo and behold,

"The Consistency of the Continuum Hypothesis, by K. Gödel."
This is the first time I have ever been aware, except from an abstract
point of view, that a place called America really exists. I have often
read about America, but it is quite different when you ask somebody
in America to send something to you, and they send it. I have been
reading the immortal work (it is only sixty pages long) alternately
with *The Magic Mountain* and find it hard to say which is the better.
[Thomas] Mann of course writes better English (or rather the transla-
tor does); on the other hand the superiority of the ideas in Gödel just
about makes up for that.

*Five years later I was drinking tea with Gödel at his home in Princeton. He
had the reputation of an inscrutable genius, but for me he was always a real and
friendly person.*

April 25, 1943

I was pleased to find here awaiting me the proofs of my first two
publications, two short notes in the *Journal of the London Mathematical Society*. I have now finally sent them to the press. Also I found an
offer of a job in the Nautical Almanac department, doing computa-
tions for various Admiralty purposes. It is no use to me now, but it is
gratifying to find that they offer you jobs like that, without your having to apply for them. On Thursday last I met Mr. Clarabut, whom
you will remember at Winchester as my chief taskmaster. He is a man
who always likes to have a finger in every pie; he works in the Foreign Office and he told us all about the big people there and how they
underrate the Japanese, while he, Clarabut, who lived in Manchuria
for some years, had been constantly telling them etc. etc. He said he
was in Cambridge in order to arrange for an honorary degree to be
given to Madame Chiang Kai-shek (not in his official capacity but
simply because he thought it ought to be done and he might as well do
it as anybody else). I do not know what his job actually is; he looked
pale and thin, so I expect he is worked hard; he must be useful in the

Foreign Office, as he speaks any number of languages (he is actually a native of Latvia) and he knows everybody who is anybody.

Clarabut had been librarian at Winchester College when I was a schoolboy and kept me busy with interesting jobs in the school library. Since the school was then 550 years old (it is now 625), the library was full of unexpected treasures.

MAY 8, 1943

Owing to a general tightening-up of the red tape, we are all (the college fire parties) being affiliated to the National Fire Service and given blue uniforms. It will not increase our duties in any way; it is just an example of the administrative mind at work. I have already filled up some forms, but with any luck I will be out of the place before the uniforms arrive. My ankle is now quite recovered, but weak from lack of use. To give it exercise, I have started an intensive programme of circumscribing the pillars in Neville's Court. The object of the game is to get round the big pillars on a ledge six inches above the ground (a) with both hands, (b) with the right hand, (c) with the left hand. (a) is quite hard, and (c) is very hard indeed; all of them are very good practice for the ankle. I have achieved (a) and very nearly (b). I had the good fortune when engaged in this pursuit to meet Littlewood, who invented the game and is the only person who can do all the pillars both ways with either hand. He explained to me at length all the finer points and was only prevented from giving a demonstration by the fact that he was wearing a dressing gown.

MAY 29, 1943

I have been seeing a good deal of Oskar Hahn, and like him very much. He introduced me the other night to Matthias Landau, the son of the great mathematician Edmund Landau, and Oskar's cousin. He is a great wei-chi expert, and we played that game till midnight, Landau needless to say winning. There is something admirable about both of them; as well as being interesting talkers, they have a sense of responsibility which is striking; I suppose that is what makes Kurt Hahn such a success. This man Landau fought for Spain in the International Brigade for a year and a half, and was interned in a foul camp in France for his pains; he said that he played more wei-chi in France than in the rest of his life. Oskar's father was the owner of a wealthy machine tool factory in Berlin, which during the 1930 slump did very well by supplying gun lathes and such to the big new armaments factories in the Urals. Although a Jew, he managed to keep his business going until 1938 and was actually asked by the government to open his works to foreign visitors to show them an example of the enlightened methods of German industry for the welfare of workers. He refused. Finally in 1938 he decided it was time to go, and he is now running a small aluminium factory in Birmingham.

The Spanish Civil War was a tragic struggle in which many English students in the years before World War II were deeply engaged. George Orwell, one of the Englishmen who fought in Spain, published in 1938 his book Homage to Catalonia, *a truthful account of the feuding and treachery on the government side that led to the victory of General Francisco Franco. During World War II we were grateful to Franco for keeping Spain neutral. As a result of Franco's neutrality, many of our bomber crewmen who survived being shot down were able to walk home through Spain and Portugal.*

Wei-chi is the Chinese version of the game that the Japanese call go. The game was invented in China and spread from there to Japan. In those days it was known in England as wei-chi, and the experts spoke scornfully of the Japanese who claimed it as their own.

JUNE 3, 1943

This morning I went for a six-mile walk before breakfast. It was a very pleasant walk; it poured with rain all the time, and I was accompanied by Oskar Hahn in his wheelchair. I am acquiring much more factual knowledge about the refugee problem now than I got from anyone else, for Oskar's mother is boss of the organisation that brought Kreisel and the others over from Germany. She is responsible for ten thousand young refugees, so she really does know something about it.

Here is a prime example of bureaucracy at its best; I have just received an order to go to the National Fire Service depot and be measured for my new fire party uniform. It was announced at the beginning of this term that the fire party was to be put into uniform; now things are beginning to get moving.

There is a two-year gap after this letter. Nothing is said about the purpose of the walk with Oskar Hahn. We walked before breakfast several times to train ourselves for a bigger walk, fifty-five miles from Cambridge to London. We both had homes in London and were finishing our time at Cambridge. Oskar had a problem with his wheelchair, which was heavy and awkward to transport by train. He asked me if I would help him to travel by road, and we came to an agreement. He would carry my heavy suitcase on his wheelchair, and I would help to push the chair when we came to steep hills. The weather was fine for the day of our big walk. We completed the trip in seventeen hours. The war gave us a unique opportunity to do it, because there was almost no traffic on the roads. With peacetime traffic, it would be impossible. We arrived at our homes without any fuss and told our parents only later what we had done. Somehow the authorities in Cambridge heard about it, and we received a letter of congratulations from Sir George Trevelyan, the historian who was then master of Trinity College.

After leaving Cambridge in 1943. I spent two years working as a civilian in the operational research section of Bomber Command. The headquarters were in a forest north of High Wycombe in Buckinghamshire. Our job was to demolish German cities and kill as many German civilians as possible. We killed about four hundred thousand, ten times as many as the Germans had killed in Britain. There are no letters from Bomber Command, since I came home to London for one day every week during the Bomber Command years. After the war in Europe ended, I stayed at Bomber Command, preparing for the move of a substantial British bomber force to Okinawa to take part in the bombing of Japan. It came as a gift from heaven when the Japanese surrendered and our trip to Okinawa was canceled. I was sud-

denly set free from Bomber Command, discharged just in time to take a position as demonstrator, the lowest grade of teacher, at Imperial College in London, where I continued to work as a mathematician while living at home. During that year I wrote one letter, while my parents were away.

DECEMBER 16, 1945

I was rung up by Oskar at ten-thirty on Friday. He spent the railway journey to Cambridge telling me all about his life in Glasgow, which he finds satisfying in spite of hardships which I never had to face at Bomber Command; exhausting physical work, the lack of a private sitting room, and a landlady who drinks whiskey. We arrived at Cambridge at four, and I failed to get a room in a hotel. Oskar was not in the least worried and said he knew just the right place for me to go, which was the house of some friends of his called Robinow. At nine o'clock we went out to the home of the Robinows, where we were welcomed tumultuously in a beautifully bohemian fashion. The household consists of Mr. Robinow who is a cytologist working in a biological research laboratory, his wife Rosie who appeared in trousers and a jersey, exotic both in voice and appearance, and finally the old lady Inona who is the mother of Rosie. Inona was an internationally famous pianist at the age of seventeen and got married and retired at about twenty-five. She was a personal friend of Brahms and used to play his later works for him as he wrote them, and she has a good collection of stories about her past. One was an account of a concert she gave in Amsterdam after coming over from England by boat. As she did not speak Dutch, it was arranged that she should be met at the boat and conveyed straight to the concert hall. The arrangements went without a hitch, the young lady was escorted to the hall, the concert was due to begin, and then for some unaccountable reason she refused to walk out onto the platform, and despite all their efforts she remained in the vestibule behind the platform and burst into tears. Fortunately Inona arrived at the hall just in time to explain to the authorities that there had been some mistake and that the wrong young lady had been met at the boat.

In the morning we had a riotous breakfast, at which we were joined by the younger generation, consisting of Antony (Nony for short) aged six and Oliver aged two. Both children are extremely good-looking, and Nony is a remarkable conversationalist. After breakfast he remarked, "Would you like to listen to some gramophone records? We've got some lovely ones of Menuhin playing Handel sonatas." And he brought out the gramophone and put on Sonata No. 4 which I was proudly able to say I had played once myself in the past. "Ah, but you didn't play it as well as Menuhin" was his reply. Afterwards he put on one of the Brandenburg concertos, and finally left me just before the end of it, remarking, "I must go to school now. Don't try to put the records away as you don't know where they go. I will do it when I come home."

At twelve-thirty I met the Hahn family. Rudolf Hahn, the father, a stout smallish man, taciturn and undistinguished-looking. The mother (Lola), tall and beautiful but with greying hair and haggard-looking eyes. The sister Benita, more English-looking than the others, having been at school at Downe House. All of them speak English with a pronounced foreign accent, and chiefly German to one another. After lunch Oskar retired with Mother to be washed and brushed up and inserted with some difficulty into his dress suit, and then we all went along together to the senate house. I do not know if you have ever been to a congregation of the senate of Cambridge University. To start with, the public orator stands up with an enormous screed of graces to be laid before the senate. In order to get through the screed in a reasonable time, he gabbles the whole thing off as fast as he possibly can, without pausing for breath or for punctuation. After the graces, the conferring of degrees began. Fortunately Oskar came near the beginning, and as soon as he was finished we made an undignified bolt for the exit.

At three we all piled into the car and set off for Banbury. During the journey I got Mrs. Hahn talking about her work. She started in 1933 in Germany organising the escape of Jews from the country, with the support of her father who was a wealthy banker and always forked out the money necessary for bribing officials, Gestapo men, etc., on a lavish scale. Having got her own children to England, she was reluctantly compelled to follow herself in 1938; however, she immediately took over a big job in the Refugee Organisation in England. All through the war she has been working like a Trojan for the organisation, driving around the country and visiting children and foster parents and local and government authorities, and since last August she has been coping with the thousand children who survived in the concentration camps and have been brought over here. The organisation caters for Jews and Gentiles alike, though most of the children are Jews. The rest of the family say that it is all they can do to stop Mrs. Hahn from adopting one of the children about once a week and from working herself to death.

Lola Hahn was born Lola Warburg, daughter of Max Warburg, and belonged to the wealthy family of bankers and art collectors who founded the Warburg Institutes in Hamburg and London. Over Oskar's desk in Cambridge was a photograph of his mother as a beautiful young woman on a horse jumping over a fence.

The house at which we arrived at six is a beautiful place, built in 1691, in the village of Cheney Middleton, Northamptonshire. When we arrived, we found a roaring coal fire in the drawing-room, and we sat round it while a sumptuous high tea was served by a real parlourmaid. The house has central heating, running hot and cold water, and two bathrooms, and a large and potentially lovely walled garden. They have been in it just six weeks, which seems almost impossible to believe as the garden is already covered with newly planted flowers, vegetables, and fruit trees (the father is a great gardener, and is particularly fond of fruit tree growing and grafting) and the whole house is furnished with lovely old furniture, Chinese paintings on silk, and spotless internal decoration. We had supper by candlelight with four excellent courses and again the impeccable parlourmaid. After supper we sat round the fire and once more roared with laughter over the antics of the officials of Cambridge University. Rudolf Hahn said that he had bought the house in response to an advertisement in the *Times*. It is not a big house, but it is roomy inside, and having two resident maids makes any house seem a palace nowadays. Finally I left them at three-thirty, with a large package of sandwiches for the train and a pot of homemade honey for my family. Of course, they keep bees as well as chickens.

Oskar inherited business skills from both his parents and rose to become chairman of the Federation of British Industries. He kept in touch with us and visited us twice in Princeton. He was a passionate sailor and loved to go sailing with his family. He died as he would have wished, dropping dead suddenly while steering his boat on a summer cruise around Scotland.

3

TRUTH AND RECONCILIATION

DURING THE SPRING of 1946 I wrote my fellowship thesis on two problems in number theory that I had solved. The main subject of the thesis was a proof of a conjecture made by Hermann Minkowski, the mathematician who had not only been a pioneer in number theory but was also one of the first to grasp the mathematical implications of Einstein's theory of relativity. The thesis was about number theory and had nothing to do with relativity. In September I returned to Cambridge as a fellow of Trinity College. During the summer of 1946 I lived with my family in London. My father sometimes invited me to lunch with him at the Royal College of Music when interesting people were there. The most memorable of these lunches was with Richard Strauss, the legendary German composer who was shunned by the international musical community after the war because he had been president of the Reichsmusikkammer, an official government organization, under the Hitler regime. My father believed strongly in reconciliation and invited Strauss to London as a gesture of friendship. As a representative of the younger generation, I helped to give Strauss a warm welcome in England. He was then over eighty years old but still vigorously alive and still writing marvelous music. He was happy to talk about the good old times before the First World War, when he was at the height of his creative powers and my father was a student in Dresden.

The letters in this chapter are mostly from Germany. In 1947 there was a meeting of students at the University of Münster. This was the first opportunity

for the German students to meet with students from other countries. Since Mün-
ster was in the British zone of occupation, the meeting was organized by the Brit-
ish military authorities. I took part, eager to meet with our enemies and hoping
to achieve reconciliation with them as my father had with Strauss. I had killed
enough Germans, and I wanted to make peace with the survivors. We all knew
that the Germans had committed atrocities, and we had too. The victims who died
in the camp at Bergen-Belsen were about as many as those who died in the fire-
storm at Dresden. I felt some personal responsibility for those who died at Dres-
den. To reach a tolerably peaceful world, we needed reconciliation more than we
needed justice.

AUGUST 8, 1947,
c/o University Education Control Officer,
HQ Mil Gov R B Münster, British Army on the Rhine

I have arrived here safely and am finding this holiday very satisfactory. Particularly I am pleased to find that I understand almost all
of the German lectures and conversation and can talk myself without as much difficulty as several members of our party. In fact, I was
astonished to find on one occasion that a German mistook me for a
German! The course is a triumph of organisation. Roughly, there are
150 students here, half from Münster university, a quarter from the
rest of Germany, and a quarter from abroad including twenty English. We live and have meals in a large, well-designed, and extremely
solid building called the Borromaeum, which is the Catholic Theological College (the students who live here in term-time are now
mostly away). It is made of reinforced concrete, even the roof is made
of concrete and faced with slate afterwards, so that it stood up to the
bombing very well. It is about two-thirds intact.

We have lectures each morning, given by all kinds of people, on
historical and social subjects mainly. Some are dull, a few very good,
and all are worthwhile from the point of view of learning the language. In the afternoons are discussions on the lectures, also sight-
seeing tours, exhibitions etc. In the evenings always some form of

entertainment, concert, film, dance or party. Last night we had a symphony concert by the Münster Municipal Orchestra (a very good orchestra with complete outfit of evening dress etc. and I should say forty or fifty members) in a concert hall seating about two thousand, which lies some way outside the town and used to be the concert hall of a German army barracks, and so escaped destruction. The main work was Bruckner's Fourth Symphony, very solid and Germanic. We had a conducted tour over the ruins of the Cathedral, with the Catholic Bishop as guide. He is a middle-aged man, very energetic and with a great sense of humour, and made the tour interesting with all kinds of anecdotes. They are beginning to rebuild the cathedral, much to the annoyance of the non-Catholics, and it is a good indication of the position of the Catholics here that this is possible. The Catholic Church, in fact, seems to be the only body capable of getting things done effectively; this course would certainly have been impossible without their help. Here in the Borromaeum, for example, we live in beautifully clean and well-furnished bedrooms, with nuns to look after them and to cook for us.

I share a room with a student of languages from Freiburg, who is a very pleasant companion. He lived in Hungary before the war, did five years in the German navy, and is now setting himself to master one fresh language every year. Since the war he has become fluent in French and English, next year comes Russian and the year after Spanish. In general, almost all the Germans seem to be equally energetic. The rations here are a share-out of British rations for the foreigners and of German rations for the Germans. As a result, the Germans consider themselves very well off; for us it is a bit of a strain, but fortunately we can slink off to the NAAFI (Navy, Army and Air Force Institutes) from time to time to supplement the diet. The life is for us, naturally, rather exhausting, but we shan't come to any harm. The German students are as one would expect, serious-minded and industrious and ready to talk about metaphysics and theology at any length; they seldom mention food difficulties, and they look, especially the girls, very far from moribund.

Tonight I have been to the Cinema, a large wooden building, in a pleasant style of architecture, one of the few which have been put up since the war. The film was *In Jenen Tagen* and was a first-rate effort, of which more will probably be heard. It was made in Hamburg, the first film to be made in the British zone since the war. As there are no film studios in this zone, it was taken entirely in the open air, but is none the worse for that. It is the story of a car, and of various people who drove it, beginning with the Jewish pianist who had to leave it behind in 1933; later it was requisitioned and went with the army to Russia; at the end it crashed into a wall during the fighting in Berlin and was abandoned, to be finally pulled to pieces by looters in search of spare parts. I went with a law student who was a wireless operator with the Luftwaffe in Russia, and he said the Russian scenes were excellent. He also talked a lot about the black market, which I intend to investigate myself one of these evenings.

I might give a rough idea of the general impression one gets from three days here. The first and most striking thing is the lack of strangeness when one walks round the town. Münster is indeed a *Trümmerstadt*; about 90 percent of the buildings in the inner town are totally destroyed; but the ruins are all overgrown with grass and vegetation and one hardly notices them. The prewar population was 120,000; during the war, because of efficient civil defense, they lost only about one thousand killed; now the population is about 90,000. The people look very like English people, the main difference being that though there are many fairly healthy looking children, there are very few babies. Trams and buses are crowded, but not impossibly so. We have had a lot of rain in the last three days (which is much welcomed as it is needed for the potato crop) but in spite of this there is nothing one could call dismal in the appearance of the town, and very little which would distinguish it from an English provincial cathedral town. Obviously one cannot tell much from such a superficial survey, and I shall perhaps change my mind in the next two weeks, but at present I have a very strong impression of Germany as a resilient and in general hopeful country. This is what the authorities want visitors to feel; on the street plan of Münster which one buys to find one's way about, is printed in large letters "Auch eine Trümmerstadt kann und soll trotz Not und allen Erschwernisse im Aussehen ihrer Strassen und Burgersteige, im Bilde ihrer Anlagen und Gärten, von der Heimatverbundenheit dem Ordnungssinn und dem Aufbauwillen ihrer Behörden und Einwöhner sprechen." It is no wonder if the French feel rather like Cato in Carthage when they come here.

The message on the street plan says, "Even a city of rubble can, in spite of losses and difficulties, speak of the love of home and the sense of order and the desire to rebuild of its leaders and its inhabitants, through the appearance of its streets and sidewalks, and the beauty of its monuments and gardens." Long sentences go better in German than in English.

AUGUST 13, 1947, BAOR, MÜNSTER

Life here continues as interestingly as it began, and I shall have a good deal to say about it. I am writing in the very comfortable lounge of the NAAFI, where we English come for seclusion as well as for sustenance. It is, of course, rather uncomfortable from the moral point of view; but the Germans well know that we have this place, and so one can only be thankful for it. You may be interested to hear what happened to me personally in this connection. When we arrived I was full of good intentions, and said, if the Germans could go through the course on the rations provided by the authorities, then so could I, and it would be good for my soul. In fact, I stuck it for four days; after that time I was so limp that I could hardly take an intelligent part in conversation, and I was sleeping for ten or eleven hours each night, and also a good deal during the day. So finally on Saturday night I swallowed my pride and a great pile of NAAFI buns; the effect was immediate, and since then I have been coming here regularly and living like a normal person. I thought it was a most useful demonstration of what this starvation means, even if I didn't carry it far. The Germans, of course, are hardened to it by now; even so, their energy under the circumstances is astounding. A few of them are invited to come here with us from time to time; but it is an invidious business, and we cannot flood the place with Germans.

On Sunday we went for a long charabanc tour of the Westfalen Wasserburgen, or moated castles. It was a lovely day, and we enjoyed it enormously. The castles themselves are picturesque, but still better was the Westphalian country and villages and woods through which we drove. It was a long day, and I got a great deal of German conversational practice as we drove along; when sitting in a charabanc and rattling over bumpy roads, one does not have to worry so much about endings! Also I spent some two hours talking Russian to a German student, and got on very well with him too; he is a Communist and is learning the language with the idea of being ready to make use of it when the time comes. (In this connection, another German remarked that this chap might have a good time while the next war lasted, but would be *entbolschewisiert* afterwards.) Unfortunately, he did not know the language as well as I, but it was good practice nevertheless. Last night we went to a performance of *Cavalleria Rusticana* by the Münster Opera Company, in the open air in the garden of the *Schloss*. The performance was not very good, but the theatre, a grassy amphitheatre overshadowed by magnificent beech and chestnut trees, and the beauty of the evening, and the silhouette of the ruined *Schloss*, amply made up for it.

On another evening we had a party at the Borromaeum at which the university quartet played Mozart quartets, and I was sitting at an open window gazing out at a landscape of ruins and shrubs and flowers and trees and twilight and feeling distinctly mystical. After that, we had performances of national songs by each national group in turn, which was interesting for the purpose of discovering who belonged to which country. We gave them "Loch Lomond" and "Green Grow the Rushes, O," and made up in vigour for what we lacked in quality. Afterwards came a succession of gradually dwindling groups, ten French, seven Dutch, six Swiss, two American, one Hungarian. Then at the end the Germans took over, and things really got going with *Studentenlieder* in large numbers.

More valuable than any of the amusements laid on for us are the conversations which occur from time to time. I will describe two of these which made a deep impression on me. First, a gathering of five men including myself; one from the Nineteenth Light Infantry (the most famous division) of the Afrika Corps, one from the motor-torpedo-boat section of the German navy in the North Sea, one from the U-boat service, one from RAF Transport Command, and one from RAF Bomber Command. We began talking about the war years, and the different views which we had of the events of that time; gradually we drifted towards personal reminiscences. I said very little, but the Germans soon became warmed to their subject and unburdened their hearts without restraint. I have seldom found the Germans so genuinely and obviously happy; a description by the U-boat sailor of what happens when a petrol tanker is torpedoed was given with the most single-minded enthusiasm. It reminded me vividly of the descriptions we used to read at Bomber Command of successful incendiary attacks, and of the elation we felt when such attacks succeeded. It is ironic that when finally enemies meet and come together as friends, they should still be able to entertain each other with such stories.

The second conversation which I wanted to mention occurred last night and was practically a monologue by a certain German Catholic theological student. He has a deep and resonant bass voice and talks in a slow and rhythmical way, with a solemn ring on the open vowels which is enough to send shivers down one's spine. He talked about God and the Church and Germany and about his own history (he piloted dive-bombers for the Luftwaffe). It was just what one usually calls German Profundity, but to meet it in the flesh was a stirring experience. He really believes, in a way which in England would be impossible, that material sufferings and adventures are of no importance in comparison with the metaphysical realities behind them. He gave a description of Russia, with its vast distances and gloomy forests, its intense loneliness and the constant terror of attack by partisans, and at the end said, Yes, Russia would make anyone into a philosopher. Naturally, not all the Germans are *tiefsinnig* like this; but it is surprising how many of them are, when they begin talking. The German character is not as mythical as I had believed.

Tomorrow we go for another long charabanc tour, this time to inspect the big Ruhr towns. Before I stop, I may as well put together a few definite conclusions which have emerged during the past week about the state of Germany. First, all the Germans agree that we must stay and occupy the country at least until the starvation is over. If we left now, they say there would at once be civil war between the towns (Communist and hungry) and the country (Christian and well-fed). Second, the general hatred and fear of Russia is quite genuine and not put on for our benefit. They believe firmly that we shall eventually be sensible enough to give them food and allow them to rebuild heavy industry so that they will be able to defend us against Russia; and, in fact, this does seem to be the best that they can hope for. Third, the democratic institutions that are being set up in this zone are held in general contempt; the outstanding problem that faces the government is to compel the country to deliver a fair share of food to the towns, and in this it has conspicuously failed. When I last wrote, I was much struck by the liveliness and energy of the people here and by the superficial normality of things. Now, after going a little further, I am struck by the deep psychological gulf which separates a country like Germany from one like England; here there is no academic and intellectual pessimism but a pessimism which permeates the people's lives.

Germany was then divided into four zones, the Northwest occupied by Britain, the Southwest by France, the Southeast by the United States, and the Northeast by the Soviet Union. The three Western zones later became West Germany, and the Soviet zone became East Germany.

AUGUST 17, 1947, BAOR, MÜNSTER

Since I last wrote, the main event has been the trip to the Ruhr on Thursday. It was arranged that we should do things thoroughly, and so we started with breakfast at six a.m. and got back to supper at eleven p.m. As it was a stiflingly hot day, we were fairly finished before the end of it; but it was very well worth it. The first place to be visited was the synthetic rubber factory at Huls, which you may remember the Americans bombed in 1943. We went over a representative selection of the works and had its functions and history explained by the members of the staff who were showing us round. The impression the place made is impossible to describe briefly; it is about forty acres in extent and consists of rectangular buildings laid out accurately, with straight roads separating them running the length and breadth of the plant; along the roads run a mass of enormous pipes and cables carrying the gases and liquids from one process to another. The buildings themselves are made of steel and concrete; there is a central control room of semicircular shape, where a man sitting at a desk in the middle has the state of each section of the works automatically displayed by electric indicators on the walls, and he can control the movements of every stage of the process by pressing various switches on his table. The whole factory was built in eighteen months, starting in the autumn of 1938. From 1940 to 1945 it produced a steady sixty thousand tons of rubber a year; the air raid of 1943 did not keep it out of action for long. It is now completely undamaged and is limited to an output of about fifteen thousand tons a year only by shortage of coal and methane gas, which are the necessary raw materials. It gave one a good idea of what the Germans can do when they get moving.

A few years later we understood why the attempt by Bomber Command to destroy the German wartime economy had failed. In 1952 Der hochrote Hahn, a book by Hans Rumpf, who had been chief of the German firefighting services throughout the war, was published. The title means "The Flaming Red Rooster." Before the war, when the plans for Bomber Command were begun, the British made a fundamental strategic mistake. We did trials to measure the effectiveness of high-explosive bombs and firebombs for destroying cities. The trials showed clearly that firebombs were five times as effective, weight for weight, as high explosives. So the decision was made to use firebombs as the primary weapon for destroying Germany. Our bombers were designed and built to carry firebombs in big quantity. High-explosive bombs were a small fraction of the payload. Our attacks were intended to destroy the German war industry by fire. Hans Rumpf understood that there was a simple way to defeat these attacks. The defense could afford to be selective. He put heavy concentrations of firefighters to protect essential military buildings and machinery and let the rest of the buildings burn. As a result, the essential factories in a destroyed city lost only about six weeks of production on the average. The workers were protected by adequate shelters, and the loss of their homes did not stop them from working. The synthetic rubber factory at Huls was a good example. It was repeatedly attacked by Bomber Command with heavy loads of firebombs and lost only a few weeks of production from start to finish. When we saw it in 1947, it looked as good as new.

The next showpiece was a Miners' Hospital at Gelsenkirchen. It is situated on a hill overlooking the town and the mines where the patients work, and is devoted entirely to the treatment of injured and sick miners. It really is a dream of what a hospital should be; a modern building, covered with creepers; all the wards facing south and overlooking gardens, with beyond a wide view of the Ruhr valley and the hills on the other side; most of the wards very small, holding two, three, or four beds, and with pictures on the walls; a large flat roof on top for sun-bathing; and a nurses' hostel in the same style alongside. The only snag that we could see is that the place is small; much too small to cope with more than a small fraction of the miners of the Ruhr district, even if you add to it another larger hospital in the same style which exists at Hamm.

After the hospital, we went to see a place called the Volksmuseum at Kettwig, where there is a collection of pictures, and we had tea. Among the pictures were some of Van Gogh's which one sees almost ad nauseam in small reproductions in England; the originals are very fine. After tea we began the drive home and should have been home earlier but for a burst tyre which gave us a pleasant opportunity to get out of the bus and watch the sun set over the Ruhr valley in the cool of the evening.

During the day we drove through Essen, Dortmund and the other most heavily bombed towns, and crossed the Ruhr valley from end to end. On the whole, it was much as I expected; the other English people were very surprised to find what a lot of open country, and very beautiful country, there is even in the Ruhr valley; the river Ruhr, where we crossed it, is a most peaceful-looking stream, winding its way over its stony bed, and sprinkled with bathers and paddling children. To me, the appearance of Essen and Dortmund came as rather a relief; I suppose I have had such a bad conscience about my Bomber Command days that I have imagined these places to be worse than they actually are. The towns are, of course, terribly destroyed, but not worse than Münster; and there is still an enormous number of modern pleasant-looking houses intact round the edges. Also, the population may be miserable, but it does not look miserable; in fact, it seemed to be wandering about the streets as good-humouredly as it does in England. No doubt the fine weather helped.

Last night we had a show in the town hall, a performance of *Ein Spiel von Tod und Liebe*, translated from the French of Romain Rolland, by the university dramatic club. It is a play about the French Revolution, very romantic and stirring. Fortunately I was near the front and could hear well, understood most of it, and enjoyed it very much. The show was given to raise money for the Ostflüchtlings-Studenten; there are a lot of these here, and they certainly are having a bad time; they are probably the hardest problem of all for this zone at present.

The Ostflüchtlingen *were the millions of Germans who were expelled from their homes in the parts of Germany that were transferred to Poland and the Soviet Union at the end of the war. The Germans and the occupying powers did a remarkable job resettling these people. They never became a permanent refugee population, as the displaced people did in a comparable situation in the Middle East.*

Today I went to the Catholic communion service, which was of a simple and intelligible kind, and we had a most able sermon from the priest in charge. Afterwards I was roped in to a meeting of representatives of various countries to discuss the formation of an international student organisation. This was a most entertaining meeting and revealed the German character to perfection; the moving spirits were all Germans, and they did all the talking; we found that before we could begin to discuss what were to be the functions of the new organisation and its relation to the many such existing organisations, we had to settle the question whether Christ was King of this present World or only the World to come; as this last question provoked a lot of heated discussion, the meeting adjourned after two hours without a vote being taken.

I am getting on well with the Germans, although I get less opportunity of speaking German now that everyone has got to know that I am English. Particularly well I get on with my roommate; he is Hungarian and so laughs with me at the Germans. With him, there is no need to discuss politics or religion, one can begin to make real friends. The best thing about him is that he is thoroughly cheerful; he lives in two attics in Stuttgart with a young wife, he has got enough coal to see him through the winter, and he does not worry about the future. His only trouble at present seems to be that he is homesick because he has never been so long away from his wife before. Richard Käferbock is his name.

A month after coming home from Germany, I was on my way to America. The Commonwealth Foundation made all the arrangements for me to travel in style. In their eyes I was not a mere student but an ambassador of goodwill between Britain and America. I was expected to behave like an ambassador and to enjoy the privileges of rank and status. Here is the last letter before my American life begins.

SEPTEMBER 15, 1947,
CUNARD WHITE STAR RMS QUEEN ELIZABETH

Tomorrow at six-thirty a.m. we dock in New York. The trip has been completely uneventful. The Atlantic has been about as lively as the Solent, and the passengers about as sober as a Buxton boardinghouse. The ship is like a good hotel and chugs along merrily at a steady thirty miles an hour. Looking out from the stern over the placid sea, the wake stretches dead straight to the horizon, and it is a very thin and well-behaved wake for so large a ship. The ship is very thin and does not seem particularly large when you are on it. Particularly if you look over the side at the point of the bow or the stern, you could not tell that you weren't on the Dover-Calais packet boat. I have made friends with two of our party; one is Mark Bonham-Carter, who is going to Chicago; the other, Marcus Cunliffe, is going to Princeton. The three of us spend a lot of time together, eating, drinking, and going to the cinema and hardly speak to anybody else. The days go by pleasantly in this company, and I fill in the gaps with long stretches of *War and Peace*. Mark is a most interesting fellow and is the tacitly appointed leader of our trio. He talks about almost any subject, usually with a lot of entertaining gossip derived from his many aristocratic friends and relations. He had an exciting war, including six months in a prison camp in Italy and one month walking four hundred miles across country to the Allied lines. He describes the latter adventure as the finest walking tour imaginable.

All my companions on this trip had distinguished later careers. Mark Bonham-Carter became a member of Parliament and ended his life as Baron Bonham-

Carter. Marcus Cunliffe settled in the United States and became a well-known historian. Andrey Vyshinsky was then Soviet delegate at the United Nations and later became foreign minister of the Soviet Union. Trygve Lie was the secretary-general of the United Nations.

I wish I could tell you more about this boat and its passengers, but I've been too lazy and occupied with the comfortable routine of life to find out. Our most distinguished passenger is Mr. Vyshinsky, but he keeps to himself, and not even Mark has any gossip about him. This lack is likely soon to be remedied, for Mark's first call when he arrives in New York is to visit a friend who is private secretary to Mr. Trygve Lie and lives at the UNO headquarters. It seems that Mark will not waste his time in the States, even if he does not learn much philosophy. England already seems remote, and no doubt will soon seem more so.

4

CORNELL STUDENT

THE LETTERS *in this chapter were all written from Cornell University. I was supported at Cornell by a Commonwealth Fellowship awarded by the Harkness Foundation. When I decided to travel to America, I had the good luck to meet Sir Geoffrey Taylor in Cambridge. Taylor was a famous scientist, expert in fluid dynamics, who had been at Los Alamos during the war. I asked him where I should go to study in America. He said without hesitation, "Cornell. That is where all the bright young people from Los Alamos went after the war." I knew nothing about Cornell, but I followed his advice.*

The choice of Cornell for my first year as a student in America was lucky in two ways. It was lucky professionally, because I fell by chance into a group of exceptionally bright people at an exceptionally exciting moment when they were attacking exceptionally important problems at the cutting edge of science. It was lucky socially, because I was living in Ithaca, a small town with a friendly atmosphere, where people helped and trusted each other, and I found there the warm welcome for which small-town America was famous. Almost at once I was a member of both communities, the university and the town.

The Cornell professors who had been at Los Alamos were Hans Bethe, head of the Theoretical Division, Robert Wilson, head of the Experimental Division, and the junior theorists Philip Morrison and Richard Feynman. Members of the British Los Alamos team whom I met at Cornell were Geoffrey Taylor and Rudolf

Peierls. Other Los Alamos colleagues who came to Cornell were Viki Weisskopf
from MIT and Robert Marshak from the University of Rochester.

The students with whom I became close friends were Ed Lennox, Leon-
ard Eyges, and Walter Macafee. All three were married. Lennox had acquired
three children with his wife Helen, a young widow whose first husband was killed
in the war.

SEPTEMBER 25, 1947

The work is turning out to be as satisfactory as I had hoped. The
system here is suited to my needs, and if I don't get something done
it will be my own fault. The arrangements are so well thought out
that I wonder nobody imported them into England. Each graduate
student is obliged to nominate a committee consisting of two or more
members of the staff. This committee is then required to supervise
his work and give him all necessary help. The chairman of the committee belongs to the student's special field and is responsible for the
student's research work. The other member or members must be chosen from different fields, and their job is to see that the student gets a
wider background of knowledge. In my case, the chairman is Bethe,
and he has given me a research problem to work on which is very
much of the sort that I need. As my secondary subject I have taken (as
is usual) experimental physics, and the second member of my committee is Professor Wilson; as a result of this I go to one course of
lectures by Wilson on experimental nuclear physics and have two
afternoons a week of modern experimental technique in the laboratories. I think this should be very good for me.

Bethe is an odd figure, large and clumsy with an exceptionally
muddy old pair of shoes. He gives the impression of being clever and
friendly but rather a caricature of a professor; he was second in command at Los Alamos, so he must be a first-rate organiser as well.

Hans Bethe was born and raised in Germany, son of a Protestant father and a
Jewish mother. His parents were separated before the rise of Hitler. In 1933 he was

dismissed from his job in Germany and found a temporary position in England. In 1935 he found a permanent position at Cornell, where he remained for the rest of his life. His father remarried and remained in Germany. After the war, Hans and his father met occasionally but were never close. Hans was loyal to America and his father to Germany.

OCTOBER 16, 1947

I am seeing a lot of Bethe, far more than I expected when I arranged this trip. He has given me a lengthy calculation to do which has considerable theoretical interest although it concerns an unobservable phenomenon. I am stimulated to work hard at it by frequent discussions with Bethe and particularly by the feeling that I am on trial. Upon my success in this job will largely depend the amount of pull I shall have when new jobs come along. My calculation is now almost finished and has given a clear answer to the problem. Bethe has six research students whom he looks after in this thorough way. This is a carefully picked bunch. He said at a seminar the other day that he considered the present situation in physics the most exciting there had been since the great days of 1925–30 [*when Bethe had been in Munich*]; this attitude makes him an ideal disseminator of knowledge and is highly contagious.

I saw Governor Dewey when he opened a new building here the other day. He is a possible Republican president next year; he was very greasy. The students here are solidly left wing and liked him no more than I did.

OCTOBER 21, 1947

Bethe has a father in Frankfurt and so is well informed about the German problem; it spite of the father, he says he would have been willing to use atomic bombs to defeat Germany, so convinced was he of the evil character of the government. He had a lot of firsthand information about the German uranium project; he is scathing over

Heisenberg's claim that the German scientists for humanitarian reasons refrained from concentrating on the bomb problem. The fact, according to Bethe, is that this was true of Hahn, the discoverer of fission, but not of Heisenberg and the others who tried very hard to make the project succeed and failed through lack of assistance from the government. I read the official report on this subject, and it more or less confirms the truth of Bethe's remarks. The fact that Heisenberg failed to foresee many of the most essential features of the bomb, in particular the importance of plutonium, seems to me important as it makes one less certain about what the Russians may have managed to do.

Last night I was reading the testimony of Professor [Philip] Morrison before the Senate Atomic Energy Committee; he is one of our leading people here and was one of the team that went to Tinian to supervise the despatch of the bombs and later to Japan to report on their effects. It was a most eloquent and inspiring testimony, he must have a great intensity of conviction on the subject; I hope I shall get to know him before long.

NOVEMBER 19, 1947

Just a brief letter before we go off to Rochester. We have every Wednesday a seminar at which somebody talks about some item of research, and from time to time this is made a joint seminar with Rochester University. I am being taken in Feynman's car, which will be great fun if we survive. Feynman is a man for whom I am developing a considerable admiration; he is the brightest of the young theoreticians here and is the first example I have met of that rare species, the native American scientist. He has developed a private version of the quantum theory, which is generally agreed to be a good piece of work and may be more helpful than the orthodox version for some problems. He is always sizzling with new ideas, most of which are more spectacular than helpful and hardly any of which get very far before some newer inspiration eclipses them. His most valuable contribution to physics is as a sustainer of morale; when he bursts into the room with his latest brain wave and proceeds to expound it with lavish sound effects and waving about of the arms, life at least is not dull.

Rudolf Peierls became a close friend of Bethe during their student days in Germany. They were both driven out of Germany by Hitler and were together for a while in England. Peierls remained in England when Bethe moved to America and found a permanent position at the University of Birmingham. In Birmingham he became one of the leaders of the British nuclear bomb project. He moved to Los Alamos as a member of the British team in 1943 and worked closely with Bethe on the final design of the Nagasaki bomb. After the war he returned to Birmingham and continued to be heavily involved in international negotiations about nuclear energy and nuclear weapons. Lord Portal was the chief of the postwar British nuclear organization. He had been chief of the Royal Air Force during the war. He was a good administrator and knew how to handle scientists.

The event of the last week has been a visit from Peierls, who has been over here on government business and stayed two nights with the Bethes before flying home. He gave a formal lecture on Monday about his own work, and has been spending the rest of the time in long discussions with Bethe and the rest of us, at which I learnt a great deal. On Monday night the Bethes gave a party in his honour, to which most of the young theoreticians were invited. When we arrived we were introduced to Henry Bethe, who is now five years old, but he was not at all impressed. The only thing he would say was "I want Dick. You told me Dick was coming," and finally he had to be sent off to bed, since Dick (alias Feynman) did not materialise. About half an hour later, Feynman burst into the room, just had time to say "so sorry I'm late. Had a brilliant idea just as I was coming over," and then dashed upstairs to console Henry. Conversation then ceased while the company listened to the joyful sounds above, sometimes taking the form of a duet and sometimes of a one-man percussion band.

After this entertainment, Peierls was urged to talk about English affairs and did so with great ability. I found this most enjoyable and decided perhaps I was not so immune as I thought from the common disease of patriotism. Peierls moves a lot on high governmental levels and had a lot to say that I did not know; he stressed that the conversion of industry to export which has been taking place in recent months would hardly begin to produce its worst effects on home supplies till next year; also he said Lord Portal is a great success as controller of atomic energy, as he does not try to interfere with things he does not understand, in contrast to certain people we could mention over here. It was surprising to me how completely Peierls understands, and obviously believes in, the English political system and such complicated phenomena as the monarchy and the House of Lords; he is very firm in his allegiance and also in support of the present government.

Peierls was shown round the new and lavishly built nuclear laboratory, which will be opened formally with a party on Saturday. Feynman remarked that it was a pity to think that after all the work that the builders had put into building it, so little would probably be done by the people who lived in it; and Bethe said that only because of the steel shortage is any nuclear physics worth mentioning done in the United States. This may well be true; the most outstanding experiments in the world are at present being done at Bristol by Powell with no apparatus more elaborate than a microscope and a photographic plate.

Cecil Powell, and his team of sharp-eyed ladies scanning photographic plates with powerful microscopes, discovered particles stopping in the plates with other particles starting where the first ones stopped. This proved that there are two kinds of mesons, later given the names pions and muons, the pions interacting strongly and the muons weakly with ordinary matter. This discovery was a big step forward in the new science of particle physics.

NOVEMBER 27, 1947

[Philip] Morrison is a young professor of theoretical physics for whom I have a considerable admiration; he is the most politically active of all the Los Alamos people here; his chief scientific achievement was the construction of the fast neutron chain reactor, a nerve-racking device which consists essentially of a lump of plutonium kept permanently in a state in which it just doesn't blow up.

The fast neutron device was built at Los Alamos and given the name Jezebel. It was important for measuring the state of affairs in a not-quite-exploding bomb.

The trip to Rochester last week was a great success. I went there and back in Dick Feynman's car with Philip Morrison, and we talked about everything from cosmic rays downward; the seminar was given by Morrison, and we then repaired to the house of [Robert] Marshak, who is the chief physicist at Rochester and is just back from Europe. Marshak is also politically active and had long talks in Europe with [Frédéric] Joliot, [Nevill] Mott, and various other people. The general impression he got was that the Europeans said, "It's up to you to make your government behave; we'll support you when we can, but there is very little we can do from this end." I get the impression that the political responsibility that has been thrust onto scientists here has been very good for them; they are almost all well-informed and liberal-minded and badly worried at the way things are going. It hits them much harder than it hits scientists in England, partly because the country itself has so much greater power in the world, but mainly because of the chaotic condition of the administration here. In England, if an important decision is to be taken, there will probably be at least some department officially responsible for it, and the scientists can get their views considered without undue difficulty; while here the situation has been so fluid that the scientists must force themselves into the picture by all means at their disposal, which means in practice sitting in Washington and badgering influential people.

Bethe has asked me to give a short talk at a meeting of the American Physical Society in New York in January on the work I have done in the last two months; this is very gratifying, even if (as I think) the work will not interest people much. It will be an excellent opportunity for meeting people. I am now just about to start on a second major problem, which should be rather more interesting, if it turns out successfully.

On Saturday we had our great inaugural party for the synchrotron building. It was a great success; I played my first game of poker and found I was rather good at it; I won thirty-five cents. The synchrotron itself does not arrive for some time yet, so the building is still empty. The party consisted chiefly of dancing and eating; Bethe and Trudy Eyges danced together for about an hour, very beautifully, while Rose Bethe and Leonard Eyges exchanged disapproving glances.

DECEMBER 7, 1947

Yesterday I had a talk with Bethe about my future. Bethe told me that unless I raise objections, he will press for me to be given a second year; he said this was "in the interests of science as well as in your own interests." He said I should spend the second year at Princeton with Oppenheimer, and that Oppenheimer would be very glad to look after me. When I first went to see Peierls at Birmingham, he told me that Oppenheimer was the deepest thinker at present in the field of physics, but at that time he was director of the physics department of the University of California and still involved in a lot of secret work at Los Alamos, so that my position would be rather dubious if I went to work with him. However, this summer Oppenheimer announced that he was fed up with all this secrecy and wanted to do some real physics, threw up the California job, and moved to Princeton. So one could hardly ask for anything better than to go to him.

Oppenheimer moved from Berkeley to Princeton in 1947. He invited a group of young physicists to the Institute for Advanced Study in Princeton with the intention of working with them. He hoped to resume his career as an active research physicist after the interruption caused by the war. This hope was never fulfilled. He remained heavily involved in government business and public affairs. He could never again give his full attention to research. As director of the institute, he helped the young physicists as a listener and a critic but never as a collaborator. He did not eat and talk with us at lunch every day as Bethe did at Cornell. He never sat down at his desk doing detailed calculations. I showed him my work to discuss it after it was finished, not while it was in progress. At the institute I learned more from my young colleagues than I did from Oppenheimer.

All this shows how fundamentally right was the idea that made me change from mathematics to physics, in spite of many discouragements. I have done nothing in the last two months that you could call clever or difficult; nothing one-tenth as hard as my fellowship thesis; but because the problems I am now dealing with are public problems and all the theoretical physicists have been racking their brains over them for ten years with negligible results, even the most modest contributions are at once publicised and applauded. If ever I should have the luck to do something clever in this field, I should have to be careful not to have my head turned.

Political argument here is dominated by the problem of atomic energy. It is betraying no secret to say that there have been newspaper reports that the atomic bomb can be increased in power by a factor of one thousand with very little increase in cost, and I know enough physics (not much is necessary) to be able to see how this is done. Knowledge of such facts as these tends to strengthen the hands of the out-and-out idealists, who say "nothing but world government can save us." The main split at present is between these people (headed by Einstein) and the compromisers (headed by Morrison and supported by Bethe) who believe that the existence of superbombs is very regrettable but does not essentially change the nature of the political problem. The compromisers, of whom I am certainly one, believe that there is very little that anyone can do at present to alter the course of history, but nothing is lost by continuing to negotiate with the Russians as honestly as we can and making compromises on as many of their demands as possible.

DECEMBER 16, 1947

The great excitement this week has been a visit from the notorious Henry Wallace. He came at the invitation of various student societies to address a mass meeting, which was held in the big concert hall in the lunch hour, and enormous crowds went to hear him. The verdict on Wallace is that he has absolutely no chance at present of a political comeback, but he is performing a most valuable service in reminding the millions of Americans who only read their newspapers that it is possible to disagree with the foreign policy of the government without necessarily being a Russian agent.

Henry Wallace had been vice president in the third term of Franklin Roosevelt's presidency, to be replaced by Harry Truman in the unfinished fourth term. He ran as leader of the Progressive Party in the presidential election of 1948, standing for a policy of peace and reconciliation with the Soviet Union.

JANUARY 2, 1948

Several of my friends are second-generation Americans, whose parents came over from Germany or Poland or Lithuania or some such place, and I am always curious to ask them questions about their parents' histories in Europe and their reasons for emigrating and their emotional backgrounds. Always I have been amazed to find that the young people know practically nothing, and apparently care little, about such matters. It is very strange when one thinks how much we have absorbed about the history and society to which our family belonged. Not that I dislike the Americans on the whole; it is probably in the long run a good thing that they live so much in the present and the future and so little in the past. The fact that they are more alone in the world than average English people probably accounts for their great spontaneous friendliness. I had heard this friendliness attributed to the size of the country and to people's loneliness in space, but I think the loneliness in time is more important.

JANUARY 9, 1948

It seems to me that one may reasonably compare the present situation of England with the position after the Napoleonic Wars. Then we were suffering from an increasing population and a static agriculture, and it took about thirty years before the industrial revolution finally pulled us out of the soup. Now we are just beginning the second industrial revolution, and it is reasonable to expect that in twenty or thirty years we shall find the results of it solving our problems again. Meanwhile we shall muddle along somehow. It is interesting that the new ideas about photosynthesis were derived from tracer chemistry, which is a by-product of nuclear physics. Many people have said that the industrial revolution produced by nuclear physics would ultimately be effected more by tracer chemistry than by nuclear power.

See Chapter 5 for my meeting with Melvin Calvin, who was doing historic experiments with tracer chemistry in California.

JANUARY 24, 1948

A new period in physics started with the Columbia University experiments last summer which for the first time contradicted the existing quantum theory outside the nucleus. The first step was taken by Bethe when he showed how the theory could be extended to explain the Columbia results. My calculations of last term were part of the detailed carrying out of this extension. Then there was another big step in November when Julian Schwinger at Cambridge, Mass., produced a formally unified theory including Bethe's work and covering the whole of nonnuclear physics. This Schwinger theory is now generally recognised as correct, at least to a much higher degree of approximation than the old theory. It is rather like the step from special to general relativity; the new theory, although in principle very much superior to the old, only gives exceedingly small differences when applied to practical problems. However, already two experiments have been suggested apart from the original Columbia experiment which will give a chance of verifying the existence of these minute effects, and the results are already in favour of the Schwinger theory. All these experiments would be impossible to do with the necessary accuracy, without the apparatus invented during the war for use in radar.

Julian Schwinger, a young professor at Harvard, had been the most brilliant student of Oppenheimer's at Berkeley before the war. Oppenheimer invited him to come as a professor to the institute, but he preferred to stay at Harvard.

Having a new theory at our disposal, the great question is: will it succeed in making sense of nuclear forces, where the old theory so lamentably failed? During the last vacation Bethe developed a programme for applying the new ideas to nuclear forces, and the job he has given me to do is to try to carry out the programme for certain types of nuclear particle. It is very exciting to be working on such a basic problem, and there is no doubt that as more and more experimental evidence comes in, we shall ultimately find the solution. Big machines are fast coming into operation in various universities, and before long we shall gradually be given more and more experimental facts.

FEBRUARY 11, 1948

On Saturday night we had a party, a cinema followed by supper at a restaurant, attended by three physicists and their wives. We munched popcorn in the cinema in true American style, and very good it was.

The film also was quite outstanding, being called *The Treasure of the Sierra Madre*. It was a story of three men who went out to the wilds of Mexico to dig for gold, armed with guns to protect themselves against bandits. The point of the story is the gradual realisation of the men, as the gold accumulated, of the power that each of them held to shoot the others and make off with the loot. Gradually they get more and more suspicious and nervous, until finally one of them goes mad and starts shooting. There is a fairly obvious application to present-day international relations. Apart from this, it was a most artistic production, using real Spanish-speaking Mexicans instead of fake ones.

I see quite a lot of Phil Morrison nowadays, as he comes to the fortnightly play reading at the Eygeses'. After the play we usually start talking, and this usually means a brilliant exposition of some recondite subject by Morrison while the rest of us listen. He has a type of mind rare among native American physicists, interested in all branches of science and especially biology, and looking on physics more as a hobby than a job. He is always astonishing me by his use in conversation of phrases that only an Englishman could understand, and which he says he picks up from reading English novels. The last play we read was *Faust*, in a new and racy translation by an American professor; I liked it very much. When Bethe returned from his travels on Monday, he found inscribed on the blackboard in his room *Grau, teuerer Freund, ist alle Theorie, / Und grün des Lebens goldner Baum*.

"Gray, dear friend, is all theory, and green the golden tree of life." I do not know whether it was Morrison who wrote these famous lines from Faust *on Bethe's blackboard. Bethe would certainly have known that they were spoken to Faust by Mephistopheles.*

FEBRUARY 24, 1948

We had recently a package of physics journals from Japan, which have caused a sensation because they have done such a lot of first-rate work in isolation from the rest of the world. At the same time as these journals came the news that Hideki Yukawa, the leading man over there, will be at Princeton next year. This will be a great opportunity for me, because very few people have ever met him in the flesh; it was he who made the prediction of the original meson in 1935. Just as I was writing this last sentence, a remarkable coincidence happened; Bethe arrived with the great tidings, obtained over the telephone from Oppenheimer, that the people at California have for the first time made a meson. This has been the purpose for which all these big machines are being built, and its achievement at last is really epoch-making. Now we shall begin to be able to know something. Everyone is tremendously elated by this. Nothing has yet been officially confirmed.

What makes the story amusing is that the people at California have been making hundreds of mesons a day for several weeks and looking furiously through their microscopes to try to find them in the photographic film; but they didn't see any because they were not developing the film properly. Fortunately [Eugenio] Lattes, one of Powell's team, happened to be on a visit to Berkeley, and he told them that the films should be soaked for half an hour instead of for four minutes, and when they did this, lo and behold, there the mesons were.

The story was amusing to physicists because the machine in Berkeley was built by Ernest Lawrence, the original inventor of the cyclotron, who also invented the "big science" style of research in particle physics. Lawrence was supreme as a machine-builder and organizer but not so good as a scientist. His competitors were amused that he won the race to produce mesons but needed the help of Lattes from Bristol to see what he had done.

Yesterday Sir Geoffrey Taylor was here, the man who first suggested Cornell to me as a suitable place to work. He came to give a talk to the seminar about the work his people have been doing at Cambridge. It was specially interesting to me because the research students who did most of the work were the Australian contingent and good friends of mine. Taylor seemed to have a high opinion of them. Like all Taylor's work, this was done in his little wind tunnel at the Cavendish, which he built himself and claims to be the smallest in the world; and the results are about ten times as accurate as anybody else's. In this connection, Taylor told me that one of the reasons he likes Cornell is that we have here the smallest cyclotron in the world.

After dinner Bethe invited me in to discuss the state of the world over a glass of whiskey, Taylor and various other notables being there. Concerning the European situation, Taylor said he saw no reason why a solution to the problem should not be found in a deliberate policy of very large-scale emigration, and that there was certainly no other solution. To which Bethe replied, "Well, we know what emigration means. You can make people emigrate if you hold a pistol at their back or starve them to death, but nothing short of that is much use."

The "European situation" here means the fear that the rising population of Europe would be unable to feed itself. This fear was particularly strong in 1948 in England and in Germany. The expected disaster never happened, partly because birth rates remained lower than expected and partly because the Green Revolution made food production higher than expected.

February 29, 1948

Walter Macafee, the negro in our room, I see a lot of as he lives by himself like me and we are often left by ourselves in the evening to go to supper. [*In those days the word* negro *was used as* African-American *is used today.*] He is a most good-natured fellow, and his only fault is that he will not often talk about serious questions. He grew up in Texas, moved up to Ohio State College to get his university training, and resolved never to go back to the South if he could help it. He had not the resources to take a doctor's degree when he graduated and was compelled to work for a number of years in a very badly paid teaching job in Ohio; there he got married and gave up hope of rising any higher. Then came the war and with it a job in a government research laboratory working on radar, with pay and conditions customary for whites. To him, this was such wealth that he was easily able to save enough during the war to get through his doctor's degree here, which he has now nearly finished. When he has his degree, he has his government job to go back to, and there is no reason why he should not live happily ever after. He is much older than the rest of us and has the handicap of his lost years. His story is an object lesson in the wastefulness of the discrimination policy.

MARCH 8, 1948

There have recently been a number of articles on political topics in the *Bulletin of the Atomic Scientists* by Oppenheimer. They are remarkable not only for the breadth of view but for the excellent prose in which they are written, and this stimulated my curiosity so that one day I got Phil Morrison talking about Oppenheimer as a personality. Morrison has known him well since he started working under him at Pasadena, their ages being then approximately twenty and thirty. He said that at that time Oppenheimer was still an exceedingly intense and aesthetic young man and divided his leisure between reading St. Thomas Aquinas in Latin and writing poetry in the style of Eliot. He came of a wealthy American family, but went to Göttingen to study and became thoroughly Europeanised; for a long time he contemplated becoming a Roman Catholic but finally didn't. To furnish his mind, he learned to read fluently in French, German, Italian, Russian, Latin, Greek, and Sanskrit. So this to some extent explains the sensitivity of his prose and the awe in which he is held by such close friends as Bethe.

Yesterday I went for a long walk in the spring sunshine with Trudy Eyges and Richard Feynman. Feynman is the young American professor, half genius and half buffoon, who keeps all physicists and their children amused with his effervescent vitality. He has, however, as I have recently learned, a great deal more to him than that, and you may be interested in his story. The part of it with which I am concerned began when he arrived at Los Alamos; there he found and fell in love with a brilliant and beautiful girl, who was tubercular and had been exiled to New Mexico in the hope of stopping the disease. When Feynman arrived, things had got so bad that the doctors gave her only a year to live, but he determined to marry her, and marry her he did; and for a year and a half, while working at full pressure on the project, he nursed her and made her days cheerful. She died just before the end of the war.

I was wrong when I wrote that Feynman found his wife Arlene in New Mexico. He married her first in a city hall on Staten Island and then took her with him to New Mexico. The story is movingly told by Feynman in the book What Do You Care What Other People Think? *(1988), the title being a quote from Arlene.*

As Feynman says, anyone who has been happily married once cannot long remain single, and so yesterday we were discussing his new problem, this time again a girl in New Mexico with whom he is desperately in love. This time the problem is not tuberculosis, but the girl is a Catholic. You can imagine all the troubles this raises, and if there is one thing Feynman could not do to save his soul, it is to become a Catholic himself. So we talked and talked and sent the sun down the sky and went on talking in the darkness. At the end of it, Feynman was no nearer to the solution of his problems, but it must have done him good to get them off his chest. I think that he will marry the girl and that it will be a success, but far be it from me to give advice to anybody on such a subject.

MARCH 15, 1948

On Friday I went to a performance of *Winterset* by the Ithaca College dramatic society. I do not know whether you know the play, which is by Maxwell Anderson, a tragedy of the deepest dye and in my opinion very fine. But perhaps more remarkable than the play was the way it was performed, which took me completely by surprise and made me revise my ideas about American education. Ithaca College is the local university, typical of the American small-town college. It naturally suffers from an inferiority complex from always having to play second fiddle to Cornell, but it has also benefited from having on its staff a number of young research students and their wives who are nominally attached to Cornell but supplement their incomes by teaching downtown. After sixty years of symbiosis the two institutions have reached a position of equilibrium, in which Cornell is the senior partner, but Ithaca College has concentrated its efforts onto excelling in one field, namely the department of music and drama. I had heard some of the concerts given by their students, and they certainly were good, but I had not before seen their dramatics. It was obvious at once that they completely outclassed anything I had seen the students do at Cambridge. The play was a difficult one, full of passionate passages which could easily fail if badly handled, and full of silences which need careful timing. Though at the back of the hall, we could hear every word that was said. The finest testimony to their acting was this, that after the first act I was told that the play was written in blank verse, as indeed it was; the delivery was so natural that I was quite unconscious of it.

My own work has taken a fresh turn as a result of the visit of [Viki] Weisskopf last week. He brought with him an account of the new Schwinger quantum theory which Schwinger had not finished when he spoke at New York. This new theory is a magnificent piece of work, so at the moment I am working through it and trying to understand it thoroughly. After this I shall be in a very good position, able to attack various important problems in physics with a correct theory while most other people are still groping. One other very interesting thing has happened recently; our Richard Feynman, who always works on his own and has his own private version of quantum theory, has been attacking the same problem as Schwinger from a different direction and has now come out with a roughly equivalent theory, reaching many of the same ideas independently. Feynman is a man whose ideas are as difficult to make contact with as Bethe's are easy; for this reason I have so far learnt much more from Bethe, but I think if I stayed here much longer, I should begin to find that it was Feynman with whom I was working more.

Victor Weisskopf was an old friend of Bethe from his student days in Germany. He was a professor at MIT, in close touch with Schwinger, who was at Harvard.

MARCH 23, 1948

Ever since 1945 I have maintained, in argument and in my convictions, that England could never fight another war against a power in control of Europe and that it was nonsensical to support any policy that envisioned such a war as a possibility. It was always possible, and indeed likely, that America and Russia would sooner or later be dragged into a war of extermination, but it seemed that the most useful thing one could do in the circumstances was to proclaim as loudly as possible a plague on both their houses. Also one could help to preserve peace by making people as aware as possible of what an atomic war would be like, although this was always liable to produce the opposite effect to what one wanted. The central point in the position was that no European could have any interest in the outcome of such a war, because he could not expect to survive it. He had a moral obligation, no doubt, to do what he could to mitigate the effects of war in his own country, but he should not expect to be able to do much.

The trouble with these beliefs is that if you try to follow them consistently, you are left in a state of paralysis. If you are faced with a situation in which war is possible or probable, there is nothing you can do. This was roughly the state I was in until yesterday. I suppose it was just because I was alone among Americans that I clung so tenaciously to my position and at the same time felt so sorry for myself.

What happened to me yesterday may be described quite simply by saying that I caught the war fever. I have seen it happen to people often before, people from whom one would never have expected a bellicose sentiment and who suddenly one day begin to talk of fighting in a matter-of-fact sort of way. Only the odd thing was that when it happened to me, I knew that it was not a fever. It came with a sudden calmness and a feeling that I was no longer afraid, a strange certainty that in facing these terrible odds without despair lies our salvation. This afternoon, with all this fresh in my mind, I happened by chance to meet a Dutch student of agriculture to whom I had never spoken more than a few words. We happened to start talking, and I soon discovered that he is an unusually well-informed and intelligent person; he seems to know about as much about England as I do, and a great deal more about the United States. He also has had very strong pacifist inclinations during the last three years. Soon we started talking about the recent war, and it turned out that he spent the greater part of it in France as a member of what we used to call "the organisation," collecting allied airmen and shipping them across Europe to fight another day; he also had a good deal to do with receiving dumps of arms by parachute and supplying them to the *maquis*. At the end of this story, he said, "And of course, in spite of my pacifist views, if the country is occupied again, I shall be back." You can imagine that I was pleased to feel that my own change of view came before, and not after, I heard this.

MARCH 29, 1948

A queer thing happened to me the other day that pleased me very much. As I may have told you, when I was at Münster last summer, there were a considerable number of pretty German girls in the party; I probably did not tell you that there was one, by name Hilde Jacobs, who was generally agreed to be the cynosure and probably the most beautiful creature I have seen anywhere. She was usually surrounded with admirers, and I never spoke to her more than an occasional *Guten morgen*. You can imagine my surprise when I received the other day, forwarded from Cambridge, a letter from this creature, and a very charming letter at that. Needless to say, I wrote back asking for more. I like to keep in touch, however tenuously, with Germany; it seems necessary to keep fresh in one's mind the life people are living there, so that one may not tire in the effort to bring these things to the attention of comfortably ignorant Americans. My new correspondent does not give much in the way of factual information; she seems to live in a private world of her own; for this one cannot blame her.

It has always seemed to me that Communism can best be understood as a return to mediaeval society and has most of the virtues as well as the vices of that society. Looking back on the political crisis of the last few weeks, one may wonder what there was to get so excited about; after all, we seem to be pretty much where we were. I think, on the contrary, that there was good reason for the excitement. During these weeks a lot of things have been done which will have the consequence of fixing the limits within which future international events will move. It was inevitable that these things should not be done without much heart-searching.

With the benefit of hindsight sixty years later, we can see more clearly the similarity between the threat of militant Communism in twentieth-century Russia and the threat of militant medievalism in sixteenth-century Spain. The United States defeated the threat from Russia as England defeated the threat from Spain, by maintaining a strong military defense and avoiding direct confrontation.

The event which caused the heart-searching in the spring of 1948 was the decision of the United States to carry out the Marshall Plan. The plan gave money to Western European states to revive their economies. The crucial economy to be revived was West Germany. This meant a switch from the postwar policy of disarming Germany to a policy of rearming Germany. It meant recognizing West Germany as an equal partner in the military defense of Western Europe. It meant giving up hope of agreement with the Soviet Union about the political future of Europe. The chief architect of the plan was George Kennan, then working in the State Department as adviser to Secretary of State George Marshall. Kennan

later became my friend and colleague as professor of history at the Institute for
Advanced Study in Princeton.

APRIL 4, 1948

I will end up with some more political notes. The revolutionary thing about the atomic bomb is not that it is so lethal but that
it is so cheap. There is a famous passage in Philip Morrison's testimony before the Senate in which he describes the air base at Tinian in the Marianas. Tinian is a small island (about one mile by two)
which became the chief base for bombers operating against Japan.
The island was first completely levelled with bulldozers, and then
six tremendous concrete runways were laid out down the length of
the island. (They could be parallel because the wind always blows in
the same direction.) When Morrison was there, the island held some
five hundred B-29s, enough to wipe out a city in one conventional
raid; to maintain them there was a permanent ground staff of twenty
thousand men, in addition to some five thousand flying crew; in the
harbour there was perpetually at least one tanker being emptied of
petrol. Tucked away at one corner of the island was Morrison's own
group; this consisted of three or four huts with a staff of twenty-five;
these with the aid of one B-29 could also wipe out a city. What Morrison stressed was that no power on earth could maintain five hundred
Tinians, by any stretch of the imagination; but one Tinian with five
hundred atomic B-29s would be not at all impossible.

The upshot of this is that when two powers both have even mod-
erate quantities of plutonium at their disposal, to have the greater
quantity is not a decisive advantage. The decisive factor in military
strength is vulnerability. And the United States is likely to remain
enormously more vulnerable to this sort of attack than Russia (let
alone Western Europe). Hence in course of time, and especially if
Russia starts building a navy, there will be increasing pressure upon
the United States to strike before it is too late. As you say, it is very
like 1914. I am an optimist too, but only in the very long run.

April 11, 1948

When I was at the conference in New York at Christmas 1947, I noticed that the building where the lectures were given was called Pupin Hall, but I did not pursue further the history of Pupin until this week, when the name caught my eye in a bookshelf in the library; out of curiosity to see who Pupin was, I looked at the book, which happened to be his autobiography. Before I knew where I was, I was deeply absorbed in it, and I thought you might be interested in what is a most remarkable story. Michael Pupin began life as a shepherd boy in a remote part of Hungary but Serb-speaking. At the age of fifteen, having had a little schooling, he decided to run away to America; and having sold his warm clothes to pay his fare, he spent a fortnight almost frozen to death crossing the Atlantic and finally arrived in New York not speaking a word of English and with five cents as his total wealth. He managed to get a job on a farm and so spent five years moving from one job to another, using the intervals to teach himself Latin and Greek and a little science. At the end of this time he considered himself educated enough to compete for a scholarship at Columbia College and astounded the examiners by knowing the first two books of the *Iliad* by heart; he got the scholarship.

At Columbia he made a brilliant career in athletics and social success, but he decided that he wanted to be a scientist, and he could not find anybody at Columbia who could teach him any real science. One day he happened to pick up James Clerk Maxwell's little book on the new electromagnetic theory, and this so captured his imagination that he decided he could not rest until he had understood it. So when he had his Columbia degree, he took a ship to England and went to Cambridge to work under Maxwell. The first thing he did was to make a call on one of the tutors at Trinity College, and he was told politely that Maxwell had been dead for four years. However, the tutor was very helpful and said that he could find him a place at Kings if he would be willing to work for the Mathematical Tripos like the other undergraduates. So for two years he worked at what he considered "infernal mathematical conundrums," until he could stand it no longer; finding time meanwhile to row for Kings and imbibe the traditions of Cambridge. At the end of this he considered himself ready to wrestle with Maxwell's electromagnetic theory and asked for a position as a physicist at the Cavendish Laboratory. He was horrified to discover that the director of the laboratory was only twenty-eight years old, the same age as himself (it was J. J. Thompson) and decided that he must start work somewhere where his mature years would be less noticeable; so he set off again for Berlin and worked under the great Helmholtz, who was a comfortable sixty-five. Here he got his Ph.D. and a wife, and so finally returned in triumph to join the staff at Columbia.

Pupin had a very keen eye for what was going on, scientifically and otherwise, in the various countries he went to, and writes a penetrating study of America and Europe in the 1880s. He describes movingly the long visits to his family which he made in the summers while he was in Europe. His hero in science was Michael Faraday. He and Faraday both had a strongly religious approach to science, idealistic rather than practical. The greater part of his life Pupin spent in a successful campaign to spread the gospel of fundamental scientific research in America. He made some useful discoveries of a practical kind in electrical engineering, but his enduring monument is the Pupin Laboratory at Columbia, which he conceived as a centre for research in the most modern and refined parts of pure physics, to do for America what the Cavendish did for England. It is sad that he did not live to see this dream come true. The three great experiments which during the last year have confirmed the new radiation theory were all carried out first in the Pupin Laboratory.

Pupin's autobiography has the title From Immigrant to Inventor *(1922). Although his chief aim in life was to foster pure science, he never claimed to be a scientist himself. He was an inventor. He was also historically important because he was a friend of Woodrow Wilson and persuaded Wilson to push for the creation of Yugoslavia after World War 1. Yugoslavia united the Serbs, Croats, and Slovenes into one country, with the Serbian king as titular leader.*

Some more interesting news about the radiation theory came this week, via Oppenheimer. At the Pocono conference last week, Schwinger expounded to the assembled savants the definitive form of the new theory, and he was universally acclaimed as the man of the hour. When Oppenheimer returned from the conference, he found on his desk a letter from Tomonaga, one of the Tokyo physicists, enclosing some manuscripts and saying that these were being sent by post because of the considerable delays of publication in Japan. And one of these manuscripts was a complete account of a theory, almost identical with the Schwinger theory, but with the development carried a good deal farther in certain directions. Oppenheimer considered this sufficiently important to justify a circular carrying the news to all members of the conference. What was perhaps even more remarkable was that Tomonaga developed the essential ideas of the theory in 1943 and only put the finishing touches to it when he heard about the Columbia experiments of 1947. Tomonaga was known before this to be one of their bright young men. I have read several good papers by him on various subjects; he is about the same age as Schwinger.

The reason that everyone is so enormously pleased with this work of Tomonaga is partly political. Long-sighted scientists are worried by the growing danger of nationalism in American science and even more in the minds of the politicians and industrialists who finance science. In the public mind, experimental science is a thing only Americans know how to do, and the fact that some theorists have had to be imported from Europe is grudgingly admitted. In this atmosphere, the new Schwinger theory is acclaimed as a demonstration that now even in theoretical physics America has nothing to learn, now for the first time has produced her own Einstein. If the scientists can say that even in this chosen field of physics America was anticipated, and indeed by a member of the much-despised race of Japanese, this will be a strong card to play against nationalistic policies. Apart from these considerations, the flowering of physics in present-day Japan is a wonderful demonstration of the resilience of the human spirit and is admired and welcomed for its own sake.

Yesterday I went to a lecture on human population and food problems, by our professor of agriculture, a very good speaker. He had a lot of things to say which I did not know, but the total picture was the usual impossibly gloomy one as far as Europe and Southeast Asia are concerned. His hope for the future was the same as mine—more science; he ended up by stating this in a rather arresting way. In 1840 Justus von Liebig started his experiments on the nutrition of plants in Germany. Before that time, no one had ever paused to ask, what is the difference between good and bad soil, and whether anything can be done about it. Since that time, as a direct result of von Liebig's work, food production over much of the world has been doubled. Scientists now are hard at work trying to answer some of the deeper questions, what makes a plant grow, and how does it photosynthesise. The implications are fairly hopeful.

June 11, 1948

The American picnic is not exactly what we understand by the term; it starts out with fried steak and salads, cooked on an open-air grille, and served with plates, forks, and other paraphernalia; this sort of thing, like the elegance of the average American home and of the women's clothes, seems to me rather a rebirth of the Victorian era, flourishing over here by virtue of the same conditions that nourished it in England. Not only in manners, but also in politics and international affairs, I often feel that Victorian England and modern America would understand each other better than either understands its contemporaries.

5

GO WEST, YOUNG MAN

MY COMMONWEALTH FUND FELLOWSHIP paid for two academic years of study and a summer of travel in between. I was encouraged to spend the summer of 1948 traveling, to see as much as possible of the United States. That meant at least one coast-to-coast trip with visits to the great universities and natural wonders of California. I made two trips to the West that summer, the first with Feynman in his car, the second alone by Greyhound bus. Between the trips, I attended the summer school in theoretical physics at the University of Michigan in Ann Arbor, where Julian Schwinger was the main lecturer. In Ann Arbor I was welcomed by two friends from Cornell: Ed Lennox, the student of Hans Bethe, and Harald Wergeland, a Norwegian physicist on sabbatical leave from his home in Trondheim.

JUNE 11, 1948, CORNELL UNIVERSITY

My plans were greatly helped by an offer of a ride across the country by Feynman. He is going to visit his (Catholic) sweetheart at Albuquerque, New Mexico, and is driving across the country starting this week; I am to keep him company on the way out, and I shall leave him and make my own way to Ann Arbor as soon as I have had enough. It should be a fine trip, and we shall have the whole world to talk about. On this visit Feynman intends to make up his mind either to marry the girl or to agree to part; most people are prepared to wager for the former alternative.

JUNE 25, 1948, SANTA FE, NEW MEXICO

Feynman originally planned to take me out west in a leisurely style, stopping and sightseeing en route and not driving too fast. However, I was never particularly hopeful that he would stick to this plan, with his sweetheart waiting for him in Albuquerque. As it turned out, we did the eighteen hundred miles from Cleveland to Albuquerque in three and a half days, and this in spite of some troubles; Feynman drove all the way, and he drives well, never taking risks but still keeping up an average of sixty-five miles per hour outside towns. It was a most enjoyable drive, and one could see most of what was to be seen of the scenery without stopping to explore; the only regret I have is that in this way I saw less of Feynman than I might have done.

We were lucky to have cool weather all the way. Each day we drove about ten hours and five hundred miles. The first was spent crossing Ohio and Indiana. We crossed the Mississippi at St. Louis at noon on the second. The Mississippi was as I had imagined it, a thick reddish-brown colour, flowing rapidly at the narrow place where the bridge is. After the Mississippi comes the Ozark country, hilly and beautiful with flowers and woods, still poor and backward economically as one could see from the dilapidated farmhouses. On the second day we crossed Illinois, Missouri, and a corner of Kansas and stopped well into Oklahoma. To me, one of the surprising things about the trip was to find how little of it was western in appearance. The country is well wooded, mixed grazing and farming, green and well watered, not unlike New York State, right to the Mississippi and beyond it as far as Oklahoma.

At St. Louis we joined U.S. Highway 66, the so-called "Main Street of America" which runs from Chicago to Los Angeles via Albuquerque. We thought that from there on would be plain sailing, as this is one of the best-marked and -maintained roads there is. However, at the end of the second day we ran into a traffic jam, and some boys told us that there were floods over the road ahead and no way through. We retreated to a town called Vinita, where with great difficulty we found lodging for the night, the town being jammed with stranded travellers. We ended up in what Feynman called a "dive," a hotel of the cheapest and most disreputable character, with a notice posted in the corridor saying "This hotel is under new management, so if you're drunk you've come to the wrong place." During the night it rained continuously, and the natives said it had been raining most of the time for more than a week.

In the morning we went on our way, the floods having subsided, until we reached a place called Sapulpa. Here we were again stopped, and when we tried to make a detour, we arrived at the water's edge, where the road disappeared into a huge lake. Returning to Sapulpa, we were fortunate in picking up a Cherokee Indian and his wife. They live on an oilfield construction camp at a place called Shawnee and had moved over for the weekend to visit some friends who had managed to secure five quarts of hooch whiskey. In this country it is illegal to sell liquor of any sort to Indians. Having spent a happy weekend getting through the whiskey, they were now on their way back to the job at Shawnee. They were able to direct us to an unpaved and indescribably muddy road, which kept to high ground and clear of the floods. In this way we came out onto a main road running westward north of U.S. 66. After a time this road too was blocked, and we had to detour still further to the north. Here the Indians left us, approximately as far from Shawnee as they were when they started.

Later on we had the bright idea of turning on the car radio, and we then picked up broadcasts from Oklahoma City and other places giving detailed stories of the floods. In this way we were able to mark on our map all the places that were under water and plan our route accordingly. The worst disaster was on U.S. 66 west of Oklahoma City, where many cars were trapped and the occupants rescued by boat, a few also being drowned. We were able finally to thread our way back to U.S. 66 a good deal further west. What made these floods remarkable is that the country around Oklahoma City is already the parched sandy rolling country of the prairie and looks as if there never had been any water there, and if there had been, it would have been sopped up at once by the sandy soil. We were sorry not to see Oklahoma City, which is said to be a unique place, a town under which a first-class oilfield was discovered in the last few years. As a result, the town and the oilfield are hopelessly mixed up, oil wells being scattered around in people's backyards, on the roads, and even inside buildings. The parts of Oklahoma and Texas that we passed through were obviously prosperous, partly due to the oil and partly to a general industrial expansion that is going on more rapidly there than anywhere else. Cattle ranching seems to be changing fast from a business to a rich man's hobby.

At the end of the third day we were in Amarillo, Texas, in the centre of the Panhandle, a treeless expanse of smoothly carved prairie. The fourth day we drove the last three hundred miles to Albuquerque before one p.m. This was the most beautiful part of the trip, though again I was surprised to find how little of it was typical New Mexico mountains. The prairie extends halfway across New Mexico, and only the last twenty miles of our journey were in mountains, the Sandia range immediately east of the Rio Grande valley in which Albuquerque lies. As we advanced into New Mexico, the prairie grew drier and drier, until a fair proportion of the vegetation was cactus, carrying at this time of year a profusion of large bloodred flowers. Coming down into Albuquerque, Feynman said he hardly recognised the place, so much has it been built up since he was there three years ago. It is a fine, spacious town of the usual American type, very little of the Spanish surviving.

Sailing into Albuquerque at the end of this odyssey, we had the misfortune to be picked up for speeding; Feynman was so excited that he did not notice the speed limit signs. So our first appointment in this romantic city of homecoming was an interview with the justice of the peace; he was a pleasant enough fellow, completely informal, and ended up by fining us ten dollars with $4.50 costs, while chatting amiably about the way the Southwest was developing. After this Feynman went off to meet his lady, and I came up by bus to Santa Fe.

All the way Feynman talked a great deal about his sweetheart, his wife Arlene who died at Albuquerque in 1945, and marriage in general. Also about Los Alamos. I came to the conclusion that he is an exceptionally well-balanced person, whose opinions are always his own and not other people's. He is very good at getting on with people, and as we came West, he altered his voice and expressions unconsciously to fit his surroundings, until he was saying "I don't know noth'n" like the rest of them.

Feynman's young lady turned him down when he arrived in Albuquerque, having attached herself in his absence to somebody else. He stayed there for only five days to make sure, then left her for good and spent the rest of the summer enjoying himself with horses in New Mexico and Nevada.

Santa Fe sits at 6,900 feet, on the edge of a vast red flat desert, underneath the Sangre de Cristo Mountains. From the mountains comes a little river, fed by snow and rain in the peaks, which runs all the year round and makes the town possible. Other rivers over the plain there are in plenty, but none of them has any water. On a spur above the town is a big stone cross, a monument to the Franciscan friars who were killed when the Indians drove the Spaniards out for twelve years (1680–92). Seeing this country, one cannot help being amazed that the Spaniards were able to colonise it at all, or that they chose to colonise it in preference to the richer lands around. It is merciless country, with not even any mountains steep enough to throw a shadow at midday. At the same time, it is remarkable how little the Spaniards really achieved during their two-hundred-years stay.

They never pacified the Indians or improved on the Indians' methods of architecture and agriculture. Although Santa Fe is proudly proclaimed as the oldest state capital in the U.S. (the governor's palace dates from 1610), it contains very few old buildings. It is a beautiful town, and a lot of it is built in the attractive adobe style, but most of the adobe houses are post-Spanish, and the style is Indian rather than Spanish. Adobe is the style of the Indian pueblos, red mud bricks and projecting wooden beams. Of the Spanish civilisation, only two things remain, the narrow winding streets and the language of the people. It is surprising to find, in a city which is completely American-built and where every road sign and advertisement and notice is in English, that three-quarters of the conversation in the streets is Spanish. Since the population of the whole state was fifty thousand in 1850 when the U.S. took it and is now half a million, a large proportion of the Spanish-speaking people must be Mexican immigrants. It was only when American railways and roads conquered first the Indians and then the deserts that the territory became habitable.

There are two kinds of Indians; the Pueblo who live in large communal settlements, which we think of now as villages but which the Spanish conquerors always referred to as cities, and the seminomads of which the chief tribe is the Navajo. The Pueblo Indians are the older and built up a high civilisation during the period 1000–1300; later they were almost annihilated by the nomads and only saved by European protection. They now still live in a few of the original pueblos, but the pueblo was designed as a fortress, and in the peaceful world of today the pueblos are mostly deserted and the inhabitants have built themselves huts scattered around on the fields they till. The nomads are now worse off, being less capable of adapting themselves to peaceful conditions; they live further from contact with civilisation, raising sheep on tracts of desert and forest, and are now suffering badly from overpopulation. All these details, and a great deal more, I learned from the excellent museum of ethnology and archaeology here. I have also wandered about the town and the neighbour-hood looking at things for myself, but one sees much more in the museum.

July 2, 1948, Chicago

From Santa Fe to here I came by bus, taking it in easy stages, via Denver, Kansas City, St. Louis. The buses that make these long trips are all run by a group of companies called Greyhound Lines and are uniform all over the country. They are a great institution, and they have beaten the railways at the passenger-carrying game; as roads improve (there is a tremendous lot of new construction going on now, as was plain to see in every state), the victory will be even more deci-sive. The buses' main advantage, which must in the long run tell, is cheapness; the railways charge half as much again for a given trip and run at a loss. Also the buses' routes can be more flexible and reach more places. Santa Fe, which is a state capital, is twenty miles from the nearest railway station. The railways at present are still quicker but not much. (When Feynman and I were driving at seventy mph along a good stretch of road between Amarillo and Albuquerque, we were overtaken by a Greyhound bus doing the same trip.) On the longer runs the buses are air-conditioned, and for night trips they are intermediate in comfort between a railway seat and a railway sleeper.

For the most part, wishing to see the country, I travelled by day and stopped for the nights. The ride from Denver to Kansas City took twenty hours and included a complete night; we traversed Kan-sas from end to end, and it was just as I had imagined it to be. (Kan-sas has always had a romantic attraction since I read *The Wizard of Oz* at a very early age.) This night in the bus turned out to be one of the best parts of the trip. (Experienced bus travellers usually sleep by day and make their social contacts at night; reading lamps are also provided.) I became involved in conversation with a boy of eighteen who was going on leave from his navy station at San Francisco to his home in Carolina, and a girl of seventeen who was going from one place in Kansas to another place in Kansas. I did most of the listening and little of the talking. The two of them were great talkers and kept it up in fine style until the sun broke through on the horizon ahead of us, ranging over love affairs, family histories, god, and politics in turn (the opposite order to that in which I should have proceeded, and thereby hangs a great part of their character); the two of them were both strongly Christian; leaders of high school religious groups and articulate in their opinions about everything. I learned more in that night about the American Way of Life, and perhaps about the way of life of people in general, than one ever learns by daylight. At times they made me feel very old, and at times very young.

One may hope one day to see these big buses, and the roads on which they run, flourish in other parts of the world where distances are large and populations scattered. During the last ten years in the U.S., the mass of the population, as opposed to the professional people or the pioneering types, has been moving about the country as never before, and this mixing up has already done much, and will do much more, to even out regional differences and fanaticisms.

It is an interesting study to observe the negroes and their houses, in St. Louis, in Chicago, and in Ypsilanti. All three towns have a large negro minority. In Chicago the negro district looks like a London working-class suburb; in St. Louis it is a great deal worse, and in Ypsilanti a great deal better. In Ypsilanti the negroes draw handsome wages from the car factories where they work.

Now, sixty-eight years later, Ypsilanti is in a sorry state. The car industry is flourishing, but the number of well-paid manufacturing jobs that it provides is sadly diminished. The workers' homes that looked so neat and clean in 1948 are now decaying slums.

So, finally, to Chicago, where I arrived on Wednesday. I am staying at a big, cheap, and very nice YMCA hotel, in the centre of the city and with one block of buildings and a few hundred yards of park between me and the lake. Since I arrived the sun has been shining and a cool breeze has been blowing in from the lake, and I feel I could very happily stay here indefinitely. There is an excellent art gallery nearby, at which I spent a long time yesterday; also bookshops, theatres, a zoo, and an endless variety of park benches. The city is noisy (they have still an elevated railway), but I am twelve floors up and not bothered by it. Above all, it is the lake which makes the city so pleasant. Because the waterfront is not taken up like New York's with docks, the lake is accessible. To walk through the centre of this city makes a very fine object lesson in the mutability of human affairs. Among the skyscrapers and the boulevards and the department stores, you come to a little monument which reads "On this spot stood Fort Dearborn, which in 1812 was attacked and besieged by British troops. The garrison, having made an escape by night, was afterwards brutally massacred, together with its women and children, by Indians." And a little further on is another monument which reads "On this spot in the year 1805 was born Helen, the first child to be born in the city of Chicago."

July 22, 1948, Ann Arbor

I have met here a graduate student called Park who was working in the U.S. Eighth Air Force Operational Research Section at High Wycombe while I was at our Bomber Command. We may even have met there, though we don't remember. This gives us a lot to talk about. Park is a most intelligent man. He has a young wife who is going this week to Philadelphia as one of the thousand-odd national delegates to the Third Party convention. She treats this as rather a joke, but still she must have worked hard to be given such a position. She is delegate for a sizeable part of Michigan. It is good that the Third Party should be preaching its ideas vigorously, even if one does not believe that they could do much were they in power.

The Progressive Party, led by Henry Wallace, had split off from the Democratic Party led by Harry Truman. It attracted young idealists like Clara Park who found Truman too belligerent. There was also a fourth party, the Dixiecrats, who were Southern Democrats fighting to preserve racial segregation. In spite of the four-way split, and to everybody's amazement, Truman won the election.

AUGUST 8, 1948, ANN ARBOR

The great event of the week was the birth of Ed Lennox's first, and Helen's fourth, child. It arrived early on Monday morning and gave very little trouble to anybody. I went to see Helen and the baby yesterday. She is in a very nice small maternity hospital, which is part of the university medical school, and she gets everything done there for a total fee of $120 because Ed is on the university staff. Helen was completely well and walking about the room. Ed is very pleased because the baby has already a nose like his, whereas the other three children still have no noses worth speaking of.

When the baby arrived and turned out to be a boy, Ed asked his daughters Nena and Caroline what he should be called, and they replied immediately, "Nicolai." So Nicolai he is. The other night I had supper with the Lennoxes and spent a happy hour fighting with Nena and Caroline. I like them very much, and what is more surprising, they seem to like me. Caroline even came back afterwards to kiss me goodnight, an entirely unexpected honour. Caroline, just five years old, is showing many signs of awakening character and intelligence. The other day when she was alone with Helen, she asked for the first time the difficult question, how it was that the first baby got born, since all babies nowadays seem to have mothers to bear them. Helen was not prepared for this and, being a Catholic (though not a very devout one), fell back on the first thing that came to mind and told her the story of Adam and Eve. Caroline listened to this intently, thought it over for a while, and then pronounced, "Well, that is a funny story."

My days here have been spent going to lectures, working at my own ideas, and reading and talking. I have met a lot of new faces, but most of the time I am with Lennox and Wergeland. The young Mrs.

Park came back from the Wallace convention. She told us all about it, and confirmed the picture drawn in the newspapers. The important question we wanted answered was how far the party is in fact Communist and how far merely liberal. She said that she found most of the delegates, like herself, young and enthusiastic and liberal and not noticeably Communist, and this was her prevailing impression of the party as a whole at first. However, toward the end of the convention, a delegate proposed a crucial amendment to the party platform, which ran, "While deploring the foreign policy of the U.S. this Party does not necessarily endorse the foreign policy of any other country," and this amendment was defeated by a majority vote. Mrs. Park said this was evidence that at least a good proportion of the delegates were Communists, or at least were making no effort to avoid being labelled as such. She continues to work for the party; she says she does not like being used by the Communists, but she prefers that to being used by their enemies. The tragedy of the Wallace party is not that it is being used by the Communists. It has now lost what was at one time a good chance of winning enough popular support to be a strong force in the land.

AUGUST 15, 1948, ANN ARBOR

After nine days in hospital Helen Lennox and the baby came home and are both doing well. Ed recently had a letter from Hans Bethe, who is enjoying himself enormously in his old haunts in Switzerland. He finds it a paradise. I think he suffers in the U.S. from the fact that he is unable to shake off the millstone of Bomb work and responsibility from his neck. It is years since he had a long and complete vacation. Also, he came to Switzerland fresh from Germany, which he found a nightmare. In Germany he was visiting his family, and a family reunion across such a gulf of years and circumstances is a psychological ordeal for all concerned.

My correspondent Hilde Jacobs at Münster has got a permit to work in Switzerland for three months this autumn. This should be a tremendous help to her, both physically and mentally; I hope soon to hear her initial reactions to it. She and I have been getting to know each other pretty well in the last six months. It all began when she wrote to me last Christmas, and I knew then at once that she wanted more than food parcels; she wanted sympathy, companionship, and above all, hope. After thinking it over, I decided that the best I could do for her was to give her all I had and let the consequences take care of themselves. So without making any commitments or promises, I have been writing to her from time to time, letters of various kinds and on various subjects, always with a certain deliberate warmth of feeling. This has been a great pleasure to me, but I think I can say also that it has been justified by its effects on her. She seems to have grown happier and more balanced in these months.

I tell you about this rarefied flirtation that has been going on across three thousand miles of ocean, so that you should know who this girl is if I ever invite her for a holiday in England. That is something that I would not do without much more thought and consideration. I think she herself expressed the dangers of the situation better than I could ever express them, by quoting Yeats:

> I would spread the cloths under your feet,
> But I am poor, and have only my dreams.
> I have spread my dreams under your feet;
> Tread softly, because you tread on my dreams.

Whether a girl, to whom English is a foreign language, could know not only how apt but how good a stanza that is, don't ask me!

The Yeats poem "Aedh Wishes for the Cloths of Heaven" (1899), has only eight lines. The first four lines are also worth quoting,

> Had I the heavens' embroidered cloths
> Enwrought with golden and silver light,

The blue and the dim and the dark cloths
Of night and light and the half light,

(continuing into the lines that Hilde quoted).

AUGUST 20, 1948, SAN FRANCISCO

I left Ann Arbor at nine on Monday morning, taking a bus direct for Chicago, and caught the westward express that evening. The route went through Illinois, Iowa, Nebraska, Wyoming, Utah, Nevada, California; one state north of my previous excursion. As far as the end of Nebraska this was the corn belt, and a bumper corn crop at that. We went through Iowa through field after field of corn, the fields not being very big but all uniformly tall, tassel-topped, and lusciously green. Between the cornfields are grazing plots where cattle are either raised or brought in from the west for fattening on the corn.

The real prairie begins only in Wyoming, and there does not last long before it fades into mountain ranch country and finally desert. The most spectacular part of the whole trip was our descent, down the Weber River valley, from the Wyoming desert into the basin of Utah. Here we were following the route of the 1847 pilgrims, which made it all the more dramatic. The Utah basin is not all fertile, but for long stretches round the edge, and in the gorges leading out of it, the rivers have been put to work and the land cultivated intensively. Only in Switzerland have I seen such painful utilisation of every scrap of mountainside. The effect of this careful husbandry, after the wastes of Wyoming, is greatly to increase one's respect for the Mormons. I think it is typical of them that on their first terrible journey across the mountains, Brigham Young attached a newly invented revolution-counter to the wheels of his oxcart so that he could navigate with greater precision.

In Salt Lake City we stopped for two hours, so I was able to take a rapid glance at the historic monuments. Of these the most interesting was a column set up in 1897 at the fiftieth anniversary of the founding of the city, with the names of 143 original settlers engraved on it.

Of the 143, twenty-seven were then still alive. Their skill in dealing with natural obstacles, Indians, and their own private quarrels was altogether exceptional. Utah now, with its mixture of small subsistence farms and industrialised cities, its non-Mormon minority of 40 percent rapidly turning into a majority, will inevitably lose much of its homogeneity and character. But it still remains a strong example in favour of Toynbee's thesis that in the long run religion counts for more in human history than any other single factor. Comparing the achievements of the settlers in Utah and California, who were building their civilisations at the same time, one feels that Utah achieved greatness while California had greatness thrust upon it.

August 26, 1948, Berkeley

Today I saw a scientific miracle which assuredly will turn the world upside down, perhaps even save our lives; one of the great discoveries which, like most such, can be understood by everybody. So I will tell you about it while it is fresh in my mind. It will do for biology what the Wilson cloud chamber did for physics. It is essentially just a gadget, like the cloud chamber which provided the essential tool for the development of nuclear physics. In a letter some months ago I spoke of the work being done at this university on the problem of photosynthesis. Today I went to a lecture by Professor Melvin Calvin, the leader of this research; he is a small, middle-aged, but very able-looking man. He was speaking about the most recent results, which were not known when I wrote before.

The cloud chamber is a box full of humid air which can be expanded rapidly. After the expansion the air is temporarily supersaturated, and the excess moisture slowly condenses into a cloud of droplets. If the box is illuminated with a flash of light and a photograph is taken immediately after the expansion, the photograph shows visible tracks made of droplets that condense around any rapidly moving nuclear particles. The visible tracks show precisely where nuclear particles are traveling through the air, and where they are occasionally deflected by nuclear interactions. The cloud chamber was invented by the Scottish physicist Charles Wilson

in 1911 and was the main tool of nuclear physicists for many years. In a similar
fashion, Melvin Calvin's invention of tracer chemistry allowed him to make visible
the rapid chemical reactions that occur in a green plant when a molecule of carbon
dioxide is absorbed from the air and converted by the energy of sunlight into sugar.

The method Calvin uses is the following. You take a large white
sheet of blotting paper; on one corner of it you place a drop of liquid,
for example the juice of crushed seaweed cells, and let it dry. Then
you fix the sheet so that it is hanging vertically with two edges horizontal, and the upper horizontal edge you fasten into a bath of some
oily liquid like phenol. The phenol slowly trickles down the sheet,
and you let it trickle for some hours. Then you take the sheet out and
dry it again. The substances contained in the original drop of juice
will be left on the sheet, but they will be carried down by the phenol
along a line along one edge. The different substances will get carried down various distances according to how great an attachment
they have for phenol. Now you fix the sheet up again, with the edge
along which the drop is spread now horizontal and uppermost (i.e.,
at right angles to its previous position), and let a second oily liquid
(not phenol) trickle down it for some hours. Then you dry the sheet a
third time. Holding the sheet in its original position, each substance
in the cell juice has been carried a certain distance down the sheet by
the phenol, and a certain distance across the sheet by the other liquid.
Now the remarkable fact is that, if you use the right sort of paper
and the right sort of liquids and the right speed of trickle, the various substances in the juice get cleanly separated from each other, and
each substance does not get smeared out much but remains as a small
splodge at some point on the sheet. These splodges cannot be seen
except for the most brightly coloured substances. But suppose that
your seaweed has been feeding on radioactive carbon dioxide (made
now in adequate quantities in a pile). Then all the substances in the
juice contain some radiocarbon, and if you put your sheet of blotting

paper against a sheet of photographic film, you get a picture with all the splodges in their respective places.

Now comes the decisive step. Suppose you have a series of samples of seaweed which are not radioactive, having been fed on ordinary carbon dioxide. The first sample you plunge into radioactive carbon dioxide for five seconds, then quickly crush the cells and extract the juice. The second sample you give ten seconds of radiocarbon, the third thirty seconds, the fourth sixty, and so on up to as long as you like. Each sample of juice is then put on blotting paper and photographed. The pictures then show, in the most direct possible way, the progress of the delicate and transitory reactions through which the radiocarbon passes as it is assimilated. After five seconds the radiocarbon is all in one or two substances, after ten seconds in four, and after thirty seconds in ten, and after a minute it has got into the splodge which contains sugar, the final product of the reactions. (This is contrary to all the old theories of photosynthesis.) Before you can interpret the pictures, you have to know which splodge is which. For this it is necessary to go through a long and laborious process, preparing every known constituent of the cells, isolating it chemically, and including in it a sufficient quantity of radiocarbon; then the isolated substance is put onto the blotting paper, and you can see where it ends up. In this way each substance can be identified with a definite position on the picture, and once the identification is made, it is done once and for all; these identifications are taking up most of the time of the workers at present, but later on this will not be necessary.

This simple technique is an immense advance, enabling people to accumulate in a few hours detailed information which before could hardly have been reached in years. Instead of carbon, you can use any other radioactive atom, and you can vary your trickling liquids to suit your conditions. The results that have been got so far are trifling in comparison with what will be done in a few years. The long-sighted people said, when nuclear energy first came on the scene, that the application to biological research would be more important than the application to power. But I doubt if anyone expected that things would get going as fast as they have. This blotting-paper-plus-radioactivity technique is revolutionary because it means that any substance can be fed to a cell and its transformations followed second by second in detail, even in quantities too small to be seen or weighed, and with substances too unstable to stand old-fashioned stewing and chemical extraction.

Melvin Calvin won the Nobel Prize for Chemistry in 1961 for this work.

SEPTEMBER 14, 1948,
17 EDWARDS PLACE, PRINCETON

Today I registered for the U.S. armed forces, as I have to do under the recent draft act. This is nothing to be alarmed about. The point of it is to catch aliens who are permanent residents of the United States. I said, if I am called up, I will join the air force and hope to get stationed in England.

After I wrote to you last, I stayed several days in Berkeley. I read the autobiographies of Jawaharlal Nehru and Oliver Lodge, both very interesting stories. On September 2 I finally boarded the bus for Chicago, going back by the same route on which I came. The Iowa corn was now standing eight feet tall, having greatly benefited by the recent heat and rain. It was still not quite ready for reaping. Out of Iowa (not a big state) comes one-tenth of the whole food output of the United States. Corn is an immensely prolific crop to grow, and you have only to look at an Iowa cornfield to feel this. It is a great pity the stuff will not grow in England.

On the third day of the journey a remarkable thing happened; going into a sort of semistupor as one does after forty-eight hours of bus riding, I began to think very hard about physics, and particularly about the rival radiation theories of Schwinger and Feynman. Gradually my thoughts grew more coherent, and before I knew where I was,

I had solved the problem that had been in the back of my mind all this year, which was to prove the equivalence of the two theories. Moreover, since each of the two theories is superior in certain features, the proof of equivalence furnished a new form of the Schwinger theory which combines the advantages of both. This piece of work is neither difficult nor particularly clever, but it is undeniably important if nobody else has done it in the meantime. I became quite excited over it when I reached Chicago and sent off a letter to Bethe announcing the triumph. I have not had time yet to write it down properly, but I am intending as soon as possible to write a formal paper and get it published. This is a tremendous piece of luck for me, coming at the time it does. I shall now encounter Oppenheimer with something to say which will interest him, and so I shall hope to gain at once some share of his attention. It is strange the way ideas come when they are needed. I remember it was the same with the idea for my Trinity Fellowship thesis.

My tremendous luck was to be the only person who had spent six months listening to Feynman expounding his new ideas at Cornell and then spent six weeks listening to Schwinger expounding his new ideas in Ann Arbor. They were both explaining the same experiments, which measure radiation interacting with atoms and electrons. But the two ways of explaining the experiments looked totally different, Feynman drawing little pictures and Schwinger writing down complicated equations. The flash of illumination on the Greyhound bus gave me the connection between the two explanations, allowing me to translate one into the other.

As a result, I had a simpler description of the explanations, combining the advantages of Schwinger and Feynman.

At the end of a week I took a bus once again, this time a thirty-hour trip, via Pittsburgh and Philadelphia. It was a fine journey, the first day being spent crossing the plains of Indiana and Ohio, which are an endless succession of rich well-tended farms and rich ill-tended industrial cities. At nightfall we crossed the Ohio River and rose rapidly into the mountains of Pennsylvania. At midnight we reached Pittsburgh, the great steel city which produces as much steel as the entire USSR. It really is impressive, with its soot and its glare of lights and furnaces. Pittsburgh is on the Ohio River, its port is Cleveland, and it belongs to the Midwest; it is only a historical accident that it is included in Pennsylvania rather than Ohio. Between it and Philadelphia lie two hundred miles of sparsely inhabited mountains. So in a desperate attempt to hold Pennsylvania together, Pittsburgh and Philadelphia were joined by the Pennsylvania Turnpike, the most ambitious road-building project in the country. We were lucky to drive along the turnpike on our route. It is a magnificent drive, through rugged forest scenery; the road is built so that no traffic ever crosses it (it all goes under or over), and in the rare places where other roads join it, they have complete cloverleaf junctions. When it comes to a big mountain, the turnpike just goes through it. The sun rose when we were halfway along the turnpike, and so we saw it at its loveliest. About Philadelphia, the city of brotherly love, there is little to be said. It is an ugly place. A short ride from Philadelphia, across rolling downs and meadows, brought us finally to Princeton, which is a pleasant little old town, entirely supported by the university, and not in the least American-looking.

The university is a large collection of buildings, built in a solid gothic style in imitation of Oxford and Cambridge, in the centre of the town. The institute is a small building in the country, about a mile from the university, built in an unpretentious and utilitarian red brick. The pleasant surprise is that it is small and intimate, a good deal smaller than the physics building at Cornell. The institute is beautifully decorated and furnished. There is a lounge with the *Times* air edition and every other important newspaper and periodical, an excellent specialist library, a tea room, and private work rooms. I glowed with reflected glory as I walked past the doors bearing the names of Einstein, Weyl, von Neumann, and Gödel. I have been given a beautiful mahogany table in a beautifully carpeted room next door to Oppenheimer, where his five young physicists are put, to be near him and one another. I have not yet met my colleagues. When I visited the institute, there were more children there than grown-ups, Dirac with his family shortly leaving for England, and various other children playing cowboys and Indians, and von Neumann looking rather vague in the midst of the confusion. If I don't do well here, it won't be the institute's fault.

The Dirac children in Princeton were his two biological daughters, roughly ten years younger than the stepchildren whom I had met in Cambridge.

Oppenheimer will not be here for about a month. At the moment he, with most of the other important people, is in England conferring with Canadian and British scientists on the subject of secrecy. The conference was held, according to the announcement, "in view of recent technical developments in atomic energy." One may speculate as to whether this may mean (a) progress in achieving uranium power plants or (b) progress in achieving superbomb explosions. In the first case one would expect some loosening of secrecy, in the second some tightening. We shall see. After this he will be at the Solvay Conference at Brussels. The Solvay Conferences are the leading international physics conferences. They are always held at Brussels, and admission is by invitation only. You will probably hear about it in the newspapers.

Tomorrow will be exactly a year since I landed. What a tremendous success the year has been! Who would have dreamed that I should be coming to Princeton with the thought not of learning but of teaching Oppenheimer about physics? I had better be careful.

6

DEMIGODS ON STILTS

"Princeton is a quaint and ceremonious village, peopled by demi-gods on stilts."

—EINSTEIN, *in a letter to the queen of Belgium after he arrived in Princeton in 1933*

THE LETTERS *in this chapter were written from the Institute for Advanced Study in Princeton, except for one from Boston.*

In 1948 Hideki Yukawa was the most famous Japanese physicist. In 1935 he had published the first field theory of nuclear forces, based on the conjectured existence of a new massive particle, the meson. During the war he remained at Kyoto as professor, not engaged in war work, teaching and taking care of students. In 1946 he started a new English-language journal, Progress in Theoretical Physics *and succeeded in publishing it in the chaotic conditions of postwar Japan. The early issues of the journal contained amazing work done by Japanese physicists during the war when they were totally isolated from the rest of the world. Yukawa mailed copies of the early issues to Oppenheimer and other leading physicists, to let them know that Japanese science was still alive. Oppenheimer invited Yukawa to the Princeton institute in 1948, and mesons were produced experimentally in Berkeley in the same year. Yukawa won the Nobel Prize in Physics in 1949 for his prediction of the meson.*

September 26, 1948

Yukawa has turned up and is most friendly and approachable. We are hoping to get him to talk to a seminar before long. Besides him, there are few eminent people but a lot of good young ones. They are all struggling to understand the Schwinger radiation theory. I have not told them that I have been struggling to supersede it; that would be bad manners. I am planning to publish my bombshell as soon as possible, preferably before Oppenheimer comes to pull it to pieces, and meanwhile say as little as possible.

September 30, 1948

After a brief visit to Cornell to collect my belongings, I settled down to work at writing up the physical theories I mentioned in the last letter. I was for five days stuck in my rooms, writing and think-ing with a concentration which nearly killed me. On the seventh day the paper was complete, and with immense satisfaction I wrote the number *52* at the bottom of the last page. While I was struggling to get these ideas into shape, I thought they were so difficult I should never make them intelligible; however, reading the paper through after it was done, it seemed so simple and clear as hardly to be worth the effort expended on it. It is impossible for me to judge at present whether the work is as great as I think it may be. All I know is, it is certainly the best thing I have done yet.

My big paper is now finished, and I have recovered from the ordeal of writing it. Meanwhile I had a letter from Bethe inviting me to spend a day at Columbia and tell the people there about the work. I am going up tomorrow morning and will probably have an all-day session discussing these problems, just as we used to in the old days at Cornell. One thing I have discovered since I wrote the paper. I was trying to keep it as short as possible, so there are questions and developments which I avoided mentioning so as to keep the argument simple. I shall now be able to write a whole string of papers going into these various points, so I shall have no lack of subject matter for my work for the next few months. To arrive at the frontiers of physics is like breaking through a crust, after which one finds plenty of room to move in a lot of directions.

Next piece of news; I have been offered the job of chief assistant at Greenwich Observatory, with excellent prospects of being Astronomer Royal by about 1965, and have decided to turn it down. I had a long letter about this from a man called Atkinson, the present senior chief assistant and heir presumptive. He gave me all possible information about the position, painting in glowing terms the scenery and architecture of Herstmonceux. It is a strange irony that this should happen; I wonder what I should have said at the age of eight if I had been told I should one day refuse such a job. You can imagine the reasons I shall give for not applying for the position. Fundamentally, I have fallen in love with the most modern part of physics and cannot now leave go of it. During the next five years, there is a gambler's chance of my doing something substantial in this field, but only if I give it a lot of my time and attention. The important thing is to use this chance while it is here. By the time I am forty, the game will be played out. I have quite a high opinion of my ability to do most things, but one thing I know I can't do, and that is to work like Einstein in isolation and produce epoch-making work. And that is what I should have to do if I were at Herstmonceux.

One thing which I must always keep in mind to prevent me from getting too conceited is that I was extraordinarily lucky over the piece of work I have just finished. The work consisted of a unification of radiation theory, combining the advantageous features of the two theories put forward by Schwinger and Feynman. It happened that I was the only young person in the world who had worked with the Schwinger theory from the beginning and had also had long personal contact with Feynman at Cornell, so I had a unique opportunity to put the two together. I should have had to be rather stupid not to have put the two together. It is for the sake of opportunities like this that I want to spend five more years poor and free rather than as a well-paid civil servant.

OCTOBER 4, 1948

On Friday I went to New York and was welcomed by Bethe, who was as friendly and informal as ever. He talked a good deal about what he had seen in Europe and all my acquaintances he had met. He said the places where most physics was being done were Zürich, Bristol, and Birmingham. At Cambridge, he said, there was still the same lot of people displaying the same masterly inactivity. One message he brought was from [Nevill] Mott at Bristol offering me a good job as lecturer there, with expectations of a professorship soon. This is certainly much more in my line than being an astronomer. However, my feeling about that too is that I shall have enough of my life being a professor and I need not be in any hurry to start.

Nevill Mott was the professor of theoretical physics at Bristol. Together with the experimentalist Cecil Powell, he made Bristol the most active center of physical research in England at that time.

Two interesting pieces of news I heard recently which show that England still can do pretty well in physics, even if it can't build such big machines as they have here. First, the blotting paper technique for analysing organic chemicals, which I wrote about recently and which is called paper chromatography, was invented and first used at the Wool Industries Research Laboratories at Leeds, in the year 1944. These people used dyes to identify the various substances on their papers; hence the name. The use of tracers for this part of the work was suggested quite early, but Berkeley having one of the best radio-chemical labs in the world has been the first to exploit it thoroughly. Second, attached to the institute here is an electronic computer project, which is aiming to be able to handle any conceivable problem. The technical side of it is under the Radio Corporation of America, and the general design is being looked after by von Neumann. The most important single technical problem in building such a machine is to make a "memory" for it, a device for storing numbers during a calculation so that they can be used later. The RCA has invented an instrument for this purpose called a selectron tube, which will hold a thousand digits and enable any particular one to be read and used in one-twenty-five-thousandth part of a second. The complete machine was to have had a memory unit with forty of these tubes. However, the project has been held up now for some time because after two years of intensive work, the RCA still could not get their tubes to work reliably. At this point there came a report from a man called Frederic Williams at Birmingham, describing a different and much simpler kind of tube which will do the same job. And within six weeks the RCA had built a tube from Williams's specification and found that it worked satisfactorily. There is no doubt that the great fault of American science is overconcentration on fashionable fields. So long as there are in England a lot of odd groups of people working on odd things, they have nothing to fear from American competition.

I must tell you about my meeting with Gödel. There is not a great deal to be said about it, since we talked mainly about mathematics and physics; he is an amusing talker, and not so pathologically shy in his home as he is at the institute. He is a little man of about forty, with a fat little Austrian wife, and they live together seeing very little of anybody and will no doubt continue to do so for many years to come since he is a permanent member of the institute. The interesting thing to me was to learn what Gödel is doing and proposes to do in the way of research. He produced during his youth two epoch-making discoveries in pure mathematics, one in 1932 and one in 1939, and since then nothing has been heard from him. With the whole of mathematics to choose from and his unrivalled talents, I was curious to know what such a man would choose to do. The answer, when it came, was completely baffling. It turns out that he has spent the last few years working in physics, in collaboration with Einstein, on problems of general relativity.

Gödel had discovered some new solutions of Einstein's equations of general relativity, describing rotating universes. The solutions known before Gödel's discovery did not rotate. Gödel invited me to his home because he wanted to know whether there was any chance that one of his rotating solutions might be true. Were the observations of the real universe accurate enough to decide whether it was rotating or not? I knew enough about observational astronomy to answer his question. The answer was negative. The observations at that time were far from being accurate enough to detect visible effects of rotation. If Gödel had asked the question today, the answer would have been different. Since the microwave background radiation has been discovered and accurately measured, we have a far more precise understanding of the dynamics of the universe, and the Gödel solutions are now definitely excluded.

I will try to explain why this is baffling. In the first place, there is no question of Gödel suffering from deterioration of intellect; he understands general relativity and its position in physics as well as anybody and knows quite well what he is about. He has found some results which will certainly be of interest to specialists in relativity. On the other hand, it is clear to most people that general relativity is one of the least promising fields that one can think of for research at the present time. The theory is from a physical point of view completely definite and completely in agreement with all experiments. Most of the work that has been done on it recently has been done by mathematicians who were interested in the mathematics of it rather than the physics, and this work was not of much value to mathematics and still less to physics. The best papers on the subject in recent years have been those of Einstein, who has done some good things, but it was generally agreed that he was continuing work on it more as a hobby for his old age than in the hope of important new discoveries.

It is the general view of physicists that the theory will remain much as it is until there are either some new experiments to upset it or a development from the direction of quantum theory to include it. In spite of all this, there is Gödel. Von Neumann, when he found himself in a similar situation, looking around for something to do worthy of his powers, went in for calculating machines in a big way. That one can understand; indeed it was a wise move.

OCTOBER 10, 1948

There is one consideration involved in this question of physics versus astronomy which I did not mention when I wrote before. It might have been supposed until recently that, as nuclear physics was a subject in which secrecy was having a seriously hampering effect on research, at least astronomy would have the advantage of being free from this. However, it turns out that the kind of fundamental nuclear physics which I am doing at present is completely free and likely to remain so, since it is quite impracticable to use it for predicting the behaviour of matter in bulk. On the other hand, it is just in the borderline between physics and astronomy that the most delicate problems involved in constructing superpowerful atomic bombs arise, and in this field I imagine are the most jealously guarded secrets. A superbomb is probably more like a nova (new star) than anything else we can see, so the time may come when telescopes become more important military weapons than piles.

The fear that I felt, of astronomy becoming secret because of the connection between exploding stars and exploding hydrogen bombs, turned out to be unwarranted. Fortunately, the tricks that we use to make hydrogen bombs explode are quite different from the tricks that Nature uses to make stars explode. Secrecy is a problem for astronomers only if they are observing man-made objects close to the earth.

On Wednesday Oppenheimer returns. The atmosphere at the institute during these last days has been rather like the first scene in *Murder in the Cathedral*, with the women of Canterbury awaiting the return of their archbishop. In passing, I may add that Eliot [*the author of* Murder in the Cathedral] is now at the institute, having been invited by Oppenheimer to come for a three-month visit and do what he likes; I have not seen him yet. During these days the secretaries have been busily cleaning and tidying up the room of the great man, where we have been making ourselves at home during his absence. Next week we are to move into a new wing of the building which is now in process of completion. (Even the institute has not escaped the disease of postwar expansion.)

I had enormous admiration for T. S. Eliot as a poet and playwright, but I was too shy to start a conversation with him when I saw him drinking tea in the institute common room. I never found out whether he saw as I did the uncanny resemblance of the opening scene of Murder in the Cathedral, *with the women of Canterbury awaiting the return of their archbishop, and the opening scene of the institute at Princeton, with the young physicists awaiting the return of their director.*

This morning the celebrated theologian Dr. Reinhold Niebuhr was preaching in the University Chapel, and I went along to hear him and see the service. I enjoyed it very much, the last time I went to church must have been on Christmas Day in New York. The singing was enthusiastic if not polished, and the building is a pleasant one, in the style of many a Cambridge college chapel. The main attraction, the sermon, did not disappoint. Niebuhr spoke on the text "Except a grain of corn fall to the ground, and die, it abideth alone; but if it die, it bringeth forth much fruit." He spoke fast, and it took all my wits to follow him. But what he had to say was worth the effort, and I found his words both stimulating and consoling; he did not hesitate to apply his remarks to political and international problems. The gist of them was, just as the individual man can save his soul only be ceasing to worry about himself and immersing his actions in some larger ends, so also we shall stand a better chance of saving our civilisation if we do not worry too much over the imminent destruction of the little bit of it to which we happen to belong. Niebuhr has the reputation of being gloomy, but I think to anyone who has faced facts squarely, his remarks were exceptionally cheerful.

Reinhold Niebuhr came in 1957 to spend a year as a visiting member of the institute. Unlike Eliot, he sat and ate lunch every day with the younger members, and we got to know him well. His table conversation was as illuminating as his sermons.

October 16, 1948

Oppenheimer is unreceptive to the new ideas in general and in particular to Feynman. Oppenheimer shocked me when he arrived by taking a semidefeatist attitude to the whole business and showing complete lack of enthusiasm for a lot of the things I consider most hopeful of fruitful advances. It is this general attitude of hesitation which I now see I shall have to fight in the next few weeks; I am sure I shall have no difficulty in the long run, and the great thing at present is to avoid antagonising people by being impatient at their conservatism.

In the afternoons I have managed to explore the country around here. It is excellent walking country, and I have met numbers of strange new birds, insects, and plants. The weather could not be better, and I hope to continue this form of exercise indefinitely. My young colleagues are unwilling to join me, as they are obsessed with the American idea that you have to work from nine to five even when the work is theoretical physics. To avoid appearing superior, I have to say that it is because of bad eyes that I do not work in the afternoons.

An interesting member of our group has recently arrived, a young man called Abraham Pais from Holland. He spent several years during the war in hiding to escape being deported and probably killed by the Germans as a Jew. He is a man of wide interests and culture and a favourite of Oppenheimer for that reason. I hope to establish a friendship with him as time goes on.

I will write a letter now to counteract the last one, which was written at a moment when I felt like Elijah in the wilderness. Since then three things have happened which have transformed the situation completely. First, on Sunday I felt so irritable that I wrote the enclosed letter to Oppenheimer. So my remarks about teaching Oppenheimer some physics came true. On Sunday night I went for a walk into a field outside the town, where the sky was unobscured by lights, and sat down on the grass to make up my mind whether I should send the letter off. After some time I had decided to do it, and then suddenly the sky was filled with the most brilliant northern lights I have ever seen. They lasted only about five minutes, but were a rich bloodred and filled half the sky. Whether the show really was staged for my benefit I doubt, but certainly it produced the same psychological effect as if it had been. I sent the letter off. On Monday I heard from the editor of the *Physical Review* that my paper has been accepted in entirety for publication in the issue of January 15. This is remarkably quick, considering that somebody had to read through the paper and referee it in that time. Since this is one of the longest papers the *Physical Review* has ever published, they might well have asked me to shorten it, which would have been a horrible task. I have much to be thankful for. Later on Monday Oppenheimer ran into me, said he was delighted with my letter and will give me opportunity to expound my views publicly next week. Finally, the same night I went uninvited to call on some of the young people at the Institute Housing Project and found they were having a party. I was very glad to join them and sat around and talked for some time, about eight of us in all. After a time they asked me what I would have to drink, and I said whiskey since that was offered. Since this episode my relations with these people have been on a much more friendly basis, so the whiskey did me a good service.

The following letter to Oppenheimer contains some technical language which nonexpert readers should skip. Translated into plain language, the letter tells Oppenheimer to listen to what Feynman has to say and stop raising silly objections. I disguised this rude message by wrapping it up in polite and diplomatic phrases.

OCTOBER 17, 1948

Dear Dr. Oppenheimer:

As I disagree rather strongly with the point of view expressed in your Solvay Report (not so much with what you say as with what you do not say), and as my own opinions are not firmly enough based for me to put them up against yours in public discussion, I decided to send you a short written memorandum. This is a statement of aims and hopes, and I would be glad if you would read it before starting on the arid details of my long paper on radiation theory.

MEMORANDUM.

I. **As a result of using both old-fashioned quantum electro-dynamics (Heisenberg-Pauli) and Feynman electrodynamics, on problems in which no divergence difficulties arise, I am convinced that the Feynman theory is considerably easier to use, understand, and teach.**

II. **Therefore I believe that a correct theory, even if radically different from our present ideas, will contain more of Feynman than of Heisenberg-Pauli.**

III. **I believe it to be probable that the Feynman theory will provide a complete fulfilment of Heisenberg's S-matrix program. The Feynman theory is essentially nothing more than a method of calculating the S-matrix for any physical system from the usual equations of electrodynamics. It appears as an experimental fact (not yet known for certain) that the S-matrix so calculated is always finite; the** divergences only appear in the part of the theory which Heisenberg would in any case reject as meaningless. This seems to me a strong indication that Heisenberg is really right, that the localisation of physical processes is the only cause of inconsistency in present physics, and that so long as all experiments are interpreted by means of the S-matrix the theory is correct.

iv. The Feynman theory exceeds the original Heisenberg program in that it does not involve any new arbitrary hypothesis such as a fundamental length.

v. I do not see any reason for supposing the Feynman method to be less applicable to meson theory than to electrodynamics. In particular I find the argument about "open" and "closed" systems of fields irrelevant.

vi. Whatever the truth of the foregoing assertions may be, we have now a theory of nuclear fields which can be developed to the point where it can be compared with experiment, and this is a challenge to be accepted with enthusiasm.

NOVEMBER 1, 1948, HOTEL AVERY, BOSTON

My physics goes on splendidly. What annoyed me in Oppenheimer's initial lethargy was not that my finished work was unappreciated, but that he was making it difficult for me or anybody else to go ahead with it. What I want to do now is to get some large-scale calculations done to apply the theory to nuclear problems, and this is too big a job for me to tackle alone. So I had to begin by selling the theory to him. As soon as he understands and believes in it, he will certainly have a great deal of useful advice and experience to offer us in applying it. Also he may be able to help me to decide what I should do next, though I am fairly determined already on a thoroughgoing attempt to prove the whole theory consistent.

After my last letter to you I decided that I needed a long weekend away from Princeton. I persuaded Cécile Morette to come with me to see Feynman at Ithaca. This was a bold step on my part, but it could not have been more successful, and the weekend was just deliriously happy. Feynman himself came to meet us at the station, after our ten-hour train journey, and was in tremendous form, bubbling over with ideas and stories and entertaining us with performances on Indian drums from New Mexico until one a.m.

Cécile Morette was the brightest of the young physicists who arrived at the institute at the same time as I did. She was the only one who quickly grasped the new ideas of Feynman. We immediately became friends. The fact that she happened to be female was irrelevant to our friendship. She was a natural leader, she understood modern mathematics better than I did, and she had a great sense of humor.

The next day, Saturday, we spent in conclave discussing physics. Feynman gave a masterly account of his theory, which kept Cécile in fits of laughter and made my talk at Princeton a pale shadow by comparison. He said he had given his copy of my paper to a graduate student to read, then asked the student if he himself ought to read it. The student said no, and Feynman accordingly wasted no time on it and continued chasing his own ideas. Feynman and I really understand each other; I know that he is the one person in the world who has nothing to learn from what I have written, and he doesn't mind telling me so. That afternoon Feynman produced more brilliant ideas per square minute than I have ever seen anywhere before or since. In the evening I mentioned that there were just two problems for which the finiteness of the theory remained to be established; both problems are well-known and feared by physicists, since many long and difficult papers running to fifty pages and more have been written about them, trying unsuccessfully to make the older theories give sensible answers to them. When I mentioned this fact, Feynman said, "We'll see about this," and proceeded to sit down and in two hours, before our eyes, obtain finite and sensible answers to both problems. It was the most amazing piece of lightning calculation I have ever witnessed, and the results prove, apart from some unforeseen complication, the consistency of the whole theory. The two problems were the scattering of light by an electric field, and the scattering of light by light.

After supper Feynman was working until three a.m. He has had a complete summer of vacation and has returned with unbelievable stores of suppressed energy. On the Sunday Feynman was up at his usual hour (nine a.m.), and we went down to the physics building, where he gave me another two-hour lecture of miscellaneous discoveries of his. One of these was a deduction of Maxwell's equations of the electromagnetic field from the basic principles of quantum theory, a thing which baffles everybody including Feynman, because it ought not to be possible. Meanwhile Cécile was at mass, being a strict Catholic. At twelve on the Sunday we started our journey home, arriving finally at two a.m. and thoroughly refreshed. Cécile assured me she had enjoyed it as much as I had. I found a surprising intensity of feeling for Ithaca, its breezy open spaces and hills and its informal society. It seemed like a place which I belonged to, full of nostalgic memories. I suppose it really is my spiritual home.

Perhaps I should tell you something about Cécile, but I hardly know where to start. No doubt I will have occasion to mention her again, so this time I will confine myself to a bare chronology. 1922, Cécile born in France, of a wealthy industrial family. 1932, father died, leaving mother and six children. Mother soon remarries. 1940, family living in Caen when German occupation starts, Cécile studying at Caen university. 1942, Cécile starts studying at Paris university, travelling to and fro weekly. 1944, Cécile caught in Paris on D-day. House in Caen destroyed, mother and one sister killed. After Germans leave, Cécile seizes one of their flats reserved for American army and installs self. Later she collects stepfather and sister and installs them too. They are still there. 1945, Cécile goes to Dublin, Institute for Advanced Study, to learn theoretical physics. 1947, Cécile moves to Copenhagen to work in Bohr's institute. 1948, Cécile comes to Princeton with institute fellowship. Future, Cécile intends to marry and have lots of children.

The town of Caen is on the Normandy coast, at the center of the region where Allied troops landed in 1944. The town was largely destroyed in two weeks of fierce fighting before the Germans were driven out.

NOVEMBER 4, 1948

First, there is this election. Everyone is immensely pleased by the result; it is clear that it will make no great difference to the way the country is governed for the next four years, but it is of great importance politically that the Democratic Party has managed to win an election without the support of the South; the worst southern states have written themselves out of the party, and the influence of the others will be much weakened. I sat up and listened to the returns coming in for about two hours; it was clear right from the start that Truman was doing much better than expected, but even so I was as surprised as everybody else by the result next morning. One has to take off one's hat to the American voters for not being bamboozled by the newspapers and the political experts.

The 1948 election had four parties contending for the presidency, the Republicans led by Robert Dewey, the Democrats led by Harry Truman, the Progressives led by Henry Wallace, and the Dixiecrats led by Strom Thurmond. Since the Democratic voters were split into three parties, it was obvious to everybody that Truman would lose and Dewey would win.

Next piece of news. I have received a letter from Professor [Isidor Isaac] Rabi, Nobel Prizeman, brains of the Columbia Radiation Laboratory which has done all the crucial experiments of the last three years, and head of the physics department at Columbia University. Content, offer of a job on his faculty. I wrote a letter back refusing the offer and explaining my circumstances, but I must confess with bitterness in my heart. Objectively, I can think of no place in the world that would be better for me than Columbia. They have the finest experimental department in the world, and it is just in the contact with experimentalists that I have most to learn; they are weak in theorists, and so I should have a free field to cultivate in building up the theoretical side, while still being close to Bethe and Oppy for consultation; finally, Columbia is a magnificent centre where everyone visiting New York stops, and is itself in easy reach of Europe. To throw away such a chance seems madness, but it seems there is no help for it. It would not matter so much were I older, but I have not yet learnt half of what America has to teach me, and it is a grim prospect to be cut off without more than rumours and months-old reports of what Feynman or Schwinger or Columbia or Berkeley is doing. Also perhaps might be added that Cécile is intending to stay here another year.

The Harkness Foundation was generously supporting me as a student in America. When I accepted their Commonwealth Fellowship, I solemnly promised to return to England for at least two years after the fellowship. The promise was intended to ensure that the fellowship should not be used as a stepping-stone for permanent emigration of English students to America. If I accepted the Columbia job, I would clearly be breaking my promise and violating the trust of the foundation.

I believe I have never written you anything about the appearance and character of Rabi. He is a tiny, impish little man, with a broad grin on his face, and is always making good-humoured jokes at somebody's expense. Where he comes from I do not know; his eyes and colour are more Mongolian than anything else. There is a celebrated song which is often sung at physical gatherings, which describes Rabi receiving the Nobel Prize. The refrain goes, "It ain't the money; it ain't the money; it's the principle of the thing." The department at Columbia may fairly be said to worship the ground he treads on, and there is a general agreement among all experimental departments outside Columbia that it is useless to try to do molecular beam experiments, because it would take anybody else about five years to arrive at where Rabi is now.

Yesterday I went to New York and spent two hours profitably with Bethe, discussing physics. He is enthusiastic about the new theory and has been using it extensively. He told me the great thing with Oppy is not to be driven frantic by him (as many people are) but to exercise calm and patience. I had come to the same conclusion myself, and I think I have been putting it into effect successfully. On Tuesday I gave the second of my seminar talks, and Oppy interrupted constantly with criticisms, some relevant and some nonsensical, so that the audience was quite bewildered; as far as I could, I went steadily on and avoided argument. The next morning, the same audience without Oppenheimer asked me to give them another seminar on the same stuff, this time without interruptions, and so I did. To me, the interruptions provided many valuable new ideas.

NOVEMBER 14, 1948

After I wrote to you from Boston, I had an unequalled opportunity of seeing the real Boston, when I went out onto Boston Common (a very small park about one-tenth the size of Hyde Park) on a night which happened by good luck to be a coincidence of Sunday, Halloween, and Election Week. The common was seething with people, talking every possible language but mostly Irish, and there were innumerable speakers standing on soapboxes in the best Hyde Park tradition, denouncing the wickedness of the world, Mr. Dewey, Mr. Truman, Mr. Stalin, and the Roman Catholic Church. I had never before seen an American crowd taking religion and politics seriously. Boston is the most European of American cities, superficially reminiscent of London. Even the slums are old and built of brick, in contrast to the typical American slums which are built of wood and corrugated iron. After two hours pleasantly spent listening to the various orators, ending with the best of them all, a little Italian who stood high up on a bandstand and periodically pointed an accusing finger at some defenceless member of his audience, exclaiming ferociously "There stands Satan"; after two hours of this I knew less than nothing about Boston politics but a lot about Boston character. The drive back from Boston was as lovely as the drive there; trees everywhere, and of a brilliance of scarlet and gold colouring such as I had never seen even in this country. Boston may be like London, but the New England countryside is nothing like England.

Oppenheimer is in California this weekend, talking to people there. He returns on Tuesday. It is no wonder he is such a nervous wreck, with all this gadding around. The wonder is rather that he manages to keep as clear-headed as he does. I have been observing rather carefully his behaviour during seminars. If one is saying, for the benefit of the rest of the audience, things that he knows already, he cannot resist hurrying one on to something else; then when one says things that he doesn't know or immediately agree with, he breaks in before the point is fully explained with acute and sometimes devastating criticisms, to which it is impossible to reply adequately even when he is wrong. If one watches him, one can see that he is moving around nervously all the time, never stops smoking, and I believe that his impatience is largely beyond his control. On Tuesday we had our fiercest public battle so far, when I criticised some unwarrantably pessimistic remarks he had made about the Schwinger theory. He came down on me like a ton of bricks and conclusively won the argument so far as the public was concerned. However, afterwards he was very friendly and even apologised to me. When life is like this, the great thing is to keep a sense of proportion and avoid becoming a nervous wreck like Oppy. So far I think I am succeeding, but you should not be surprised when I write melancholic letters occasionally.

Fortunately there are occasional diversions. On Friday a carload of us young people drove to a local cinema where we saw *Rope*, a new film. It is a masterpiece, directed by Alfred Hitchcock in his best style. After the film, we got back into the car and hardly had we driven a block when we saw a genuine corpse, with an ugly wound on its head, stretched out across the pavement. Pais, who owns the car, drove off to fetch a doctor while the rest of us joined the gathering crowd around the corpse. The wife of the corpse turned up, not at all perturbed. She shook the corpse vigorously, and it opened its eyes and snorted something incomprehensible at her. "I told him he shouldn't go out," she said to the crowd. "He was dead drunk when he left the house. Mind you, this is not the first time this has happened." At this point a police car arrived and carried the corpse away to hospital, and the crowd dispersed. Several minutes later Pais returned, fortunately having failed to find a doctor. After this episode we assembled to drink wine in Pais's rooms.

Norman Kroll is the most mature and quiet of the young people and is married to an equally quiet biologist wife; I met them at Ann Arbor in the summer, and I like them very much though I do not see much of them now. Norman has done during the last two months some work which I consider first rate, on the Schwinger theory. I am hoping that as soon as I can get through with my series of talks to the seminar, he will take over and talk about what he has done. However, I go ahead so slowly under Oppenheimer's fire that an end is hardly yet in sight. Kroll is wisely not pressing his claims. Kroll was a pupil of Lamb at Columbia, but this is the first important thing he has done. I shall be surprised if Rabi does not now invite him back to Columbia to fill the place he offered me. Kroll deserves it as much as I do, and I shall not grudge it to him. It is pleasant to find that Kroll has been quietly going ahead with the theory and making progress, undeterred by the floods of scepticism which Oppenheimer pours on the whole business. This has a good effect on the morale of the young people, who were getting thoroughly confused and discouraged, in spite of all I could say.

Rabi did offer the job at Columbia to Kroll, and Kroll accepted it. Kroll stayed for many years as a mainstay of the Columbia physics department after Rabi retired. Kroll was particularly well suited to be in the Columbia department, since he had been a leader in the development of high-frequency radar equipment during the war. He could talk as an expert with experimenters as well as with theorists. In that way, he was better qualified for the Columbia job than I was.

Cécile amused us all yesterday by bringing down a French millionaire to see the institute (an industrial magnate of some kind). She said she hinted to him strongly that France could do with an institute of a similar sort; she said if she were made director of the French institute, she would invite all of us to come and lecture there. It will be interesting to see if anything comes of it.

The millionaire that Cécile brought to the institute was Léon Motchane, a man of many talents. He had achieved fame in three separate careers, as a mathematician, as an entrepreneur, and as an active leader of the French resistance. Cécile's plan, that Motchane could be cajoled into building a comparable institution in France, succeeded brilliantly. Within a few years, Motchane had helped to found two institutions in France, the Institut des Hautes Études Scientifiques in Bures-sur-Yvette with himself as director, and the Les Houches summer school with Cécile as director. The IHÉS is a smaller version of the Princeton institute, housed in a beautiful château in the village of Bures on the southern edge of Paris. The Les Houches summer school is a high-level school of theoretical physics, meeting for six weeks every summer in the French Alps, attracting first-rate teachers and students from all over the world. Both institutions are still flourishing sixty years later. Cécile kept her promise and invited me to teach at Les Houches in 1954. I taught the brightest class of students that I ever encountered. The brightest of the bright was Georges Charpak, who won a Nobel Prize for Physics in 1992.

November 21, 1948

During the past week I at last began to make some progress in explaining my ideas. Up till this week I had given only two talks in two months, and those two were mainly occupied with fighting Oppenheimer; several times I had been scheduled to talk and then put off at the last moment because something else had to be discussed; you can imagine how frustrating it was to go on for two months like this. However, at last on Wednesday of this week, Bethe came to my rescue. He came down to talk to the seminar about some calculations he has been doing with the Feynman theory. He was received in the style to which I am accustomed, with incessant interruptions and confused babbling of voices, and had great difficulty in making even his main points clear; while this was going on he stood very calmly and said nothing, only grinned at me as if to say, "Now I see what you are up against." After that he began to make openings for me, saying in answer to a question, "Well, I have no doubt Dyson will have told you all about that," at which point I was not slow to say in as deliberate a tone as possible, "I am afraid I have not got to that yet." Finally Bethe made a peroration in which he said explicitly that the Feynman theory is much the best theory and that people must learn it if they want to avoid talking nonsense; things which I have been saying for a long time but in vain.

After the seminar Bethe had supper with the Oppenheimers; I did not see him except during the seminar. But the next morning I found that my triumph was complete; three extra seminars had been arranged for me in one week. And in the first of these on Thursday Oppenheimer actually listened to me and did not interrupt. The next two are on Monday and Tuesday, and that will be enough for me to get the main essentials done with. Bethe is a great and good man, and I wrote to him and told him so. The tact and strategy which he used, to pull the opinion of the institute onto my side, could not have been more effective. My own researches meanwhile go on with renewed momentum. I have already enough material for a new paper to the

Physical Review. I am in no hurry to write it, and I am taking things as easy as possible. As I expected at the time, the result of writing the first paper so quickly is that there is very little of it that I should not like to change now. It was right to publish it fast for the benefit of people in Europe and Japan.

It is strange how easily and unintentionally I have slid into the position of a pundit. Now when I give a seminar, I usually find that [Eugene] Wigner has come over from the university to listen to it, and he has once come up to me afterwards to talk. Wigner is the chief professor of theoretical physics at the university, a man with a tremendous reputation and perhaps the greatest living expert on the theory of piles. He is a Hungarian by birth and a very polished and pleasant personality. It is strange that my modest contribution to physics should bring me such a reward as this. It seems somehow out of proportion that to clear up a minor muddle in one branch of physics should be such a serious matter. What is even more strange is that I find myself giving these seminars, without notes or preparation, with Wigner and such people in the audience, and without feeling nervous. A year ago this would have been completely unthinkable.

The word pile *was then used for the objects which we now call nuclear reactors. Enrico Fermi built the first nuclear reactor in Chicago in 1942, with Wigner working out the theory of its operation. Wigner then used the theory to design the first high-power reactor that was used to produce plutonium for the Nagasaki bomb. Reactor engineers all over the world have used Wigner's theory as the basis of their designs. For Wigner, the theory of reactors was only a sideline. He had broad interests in physics and in public affairs. He was one of the group of brilliant Hungarian physicists who came to America in the 1930s. He came as an immigrant, not as a refugee. His main contribution to science was to understand the laws of nature as consequences of mathematical symmetry. He showed how general ideas of symmetry could lead to detailed understanding of the behavior of atoms and nuclei.*

7

NOLO CONTENDERE

FOUR WOMEN and ten men were young physicist members at the Institute for Advanced Study for the academic year 1948–49. That was a record year for female members. The 28 percent fraction of female physicists has never been equalled in the subsequent sixty-seven years. The physicists at the institute have always tried hard to increase the proportion of women but without success. Those who arrived in 1948 had the advantage of growing up during the war when most of the young men were away. They were outstanding in quality as well as quantity. Cécile Morette from France played a leading role in the group. Sheila Power came from the Dublin Institute for Advanced Study, Bruria Kaufman from Columbia University, and Cheng Shu Chang from the University of Michigan. Bruria Kaufman had the distinction of being assistant to Albert Einstein. She published several papers with Einstein, besides other more memorable papers that she published alone. Cécile Morette stayed on at the institute for the year 1949–50, when Bryce de Witt was also a member. She and Bryce were married in 1951. Both had distinguished careers as physicists, and they raised four daughters.

NOVEMBER 25, 1948

My last week has been occupied with the three extra seminars which I had been allotted. I was anxious to finish off the whole theory in these three sessions. Fortunately the audience was also anxious to finish it off, and Oppy was cooperative. It was a hard struggle to get everything covered, but by filling the blackboards with formulae before I started, I had all the main ideas fairly put across. At the end Oppy made a short speech: "It is not possible to say on the basis of these talks that the consistency of the theory is proved, but at least we have all learnt a great deal, and shall have plenty to argue about from now on." Whereupon the exhausted audience, in no mood to start an argument, quietly departed. I came away from the last talk with a feeling of tremendous relief, my head for the first time for weeks entirely empty of ideas clamouring for expression. It will be possible now to relax, to return to the position of an ordinary young member of the institute, and to go to seminars without having to say anything. Next week Yukawa is to give the seminar, and the week after that Kroll.

The day after the last of my talks, I found in my mailbox a little handwritten note saying, "Nolo contendere. R.O." This was a typically erudite statement from Oppenheimer, telling me that he accepted my arguments. It is the Latin phrase used by lawyers to say that they do not dispute an opinion. It was his formal notice of surrender.

Last night the Oppenheimers gave their Thanksgiving party, a stand-up supper for about one hundred guests, mostly the institute and its wives. The party was quite enjoyable as such things go. The young physicists kept pretty much to themselves, and I did not speak to many people outside our circle. There were, however, two exceptions to this rule. First, it was a farewell party for T. S. Eliot, who is returning to Europe to receive his Nobel Prize and go home. Most of the time he was surrounded with elderly and distinguished people in a small drawing room apart from the main crowd; our physicist group was in the main room lamenting the fact that none of us had been brazen enough to go and talk to the great man. This conversation immediately fired the light of ambition in Cécile's eyes, and she said,

"Well, you are a lot of cowards; I'll go and fetch him out for you." So she went into the little room where the elderly and distinguished people were, came out a few seconds later with a grinning Eliot in tow, and introduced us to him one by one. After this there was a brief period of rather embarrassed conversation, which was made easier by the fact that Eliot has a sense of humour and some experience in dealing with such situations. Then Cécile returned him to the little room.

All this time Oppy was rushing around, resplendent in black tie and dinner jacket, making sure he met and spoke to everybody. He is a first-rate host and looked happier than I have seen him ever. When he spoke to me, it was to give me the recipe for some delicious Mexican savories that were being served with the supper. (He is an expert cook.) Then he rushed off to the next conversation, which might have been on any subject from football to cuneiform texts. This kind of evening is probably the nearest he ever gets to being relaxed. Mrs. Oppy I also met. She is quick-witted on a much more human level. She struck me as a friendly, direct young person, with no airs and entirely unspoiled by greatness.

The queerest and maddest part of the evening came at the end. People were then trying ineffectively to dance in the constricted space available. I was suddenly seized upon by an absurd and very drunken little woman, who ordered me to dance with her. As she is a pathetic-looking creature with a disfiguring scar on her face, I could not decently reject her. So I danced around with her for about twenty minutes, she evidently not minding how badly I danced. At the end of this she was getting so wild and jumping about so that it made me very uncomfortable, and I finally succeeded in returning her to her husband. The husband, who is a solemn and frightened-looking little man, was standing around by himself miserably while all this went on. He did not seem to talk to anybody all the evening. It makes me feel sick just to think of the horror of the lives these two people may be living. Evidently the reason the wife seized upon me for a partner is that I am the only one of the young men at the party whom she had met before. The name of the husband (I wonder if you guessed it) is Kurt Gödel.

The horror of this scene was real, but Adele Gödel was rarely drunk, and she was a good wife for Kurt when she was sober. She took good care of him and gave him what he needed, a quiet home where he could work and think in peace. When I saw them together at their home, she kept the guests comfortably supplied with tea and cake while Kurt led the conversation.

NOVEMBER 28, 1948

Yesterday a Chinese friend called Ning Hu was here from Cornell, and he and I and Sheila Power went for a long walk. (Ning Hu is a friend of Sheila since he was two years with her in Dublin.) It was warm and sunny and breezy like an English May, and we walked a good twelve miles round Carnegie Lake where Princeton University does its rowing. We talked about Ireland and about China and its problems. Afterwards Ning Hu came to my room, and we talked some more. I was delighted to get to know him. He is lonely and very glad of the company. He is depressed at the state of affairs in China, and even more depressed at the lack of sympathy with which these affairs are talked about by Americans and Chinese in this country. He comes from the generation which grew up in the years 1925–35 and which has produced all the well-known Chinese scientists; for that brief period of years, the Chinese universities began to stir from their slumbers and made astonishing progress in catching up with the rest of the world. The students and young professors were energetic and enthusiastic and had enough material goods to relieve them from constant preoccupation with making a living. Then after 1935 he watched the hard-won progress gradually destroyed; he moved with the university from Peiping to Kunming in the far Southwest and tried to build things up again there; but one by one the bright stars either gave up the struggle or fled in desperation to Europe or the U.S. He himself came to the U.S. in 1941, and he is forced to stay here at least for several years, since only by earning U.S. dollars can he hope to keep his parents alive in China. Like all the people I have met from China, he is no Communist but believes that the only hope now is a complete Communist victory as soon as possible. I hope to see more of him and hear more of his story in the next few days.

That was the last year of the Chinese civil war, when the victorious army of Mao Zedong was defeating the disorganized government forces of Chiang Kai-shek. China, ravaged by years of Japanese occupation and years of civil war, was in desperate need of peace. One year later Chiang with the remnants of his army was driven into exile in Taiwan, and Mao was in Beijing proclaiming the birth of the People's Republic of China. Ning Hu, well aware of the political risks and material hardships that awaited him, decided to return to China. He returned in 1951 and was well treated by the Communist regime. He enjoyed a distinguished career as teacher and researcher until his death in 1997.

Sheila Power is also a person I am glad to get to know. She is a Catholic like Cécile, but in every other respect as unlike as possible, quiet and unassuming and un-formidable. Cécile now seems to have decided that ordinary buses and trains are much too slow for her; this weekend she is spending at Chicago, flying there and back, and at Christmas she is to fly to Denver, Colorado, for a week's skiing holiday. Fortunately she always seems to have plenty of money.

I enjoyed having Cécile as a colleague and companion but had no wish to become involved with her personally. The word that best described her was formidable, either in English or French. We liked to make jokes about her role as a modern version of Joan of Arc, the young girl who put on soldier's clothes and led the French army to defeat the English at Orléans in 1429. Cécile herself accepted this role and later used her summer school at Les Houches to lead a revolution in the teaching of physics in France.

December 4, 1948

We are all watching anxiously to see if the Soviet physicists are in for serious trouble. The recent newspaper attacks on them are nothing new and are not necessarily indicating anything more than the idiotic arguments that have been going on for twenty years or more. However, in view of what happened to the biologists, the outlook is pretty black. The issue in physics is curiously parallel to the issue in biology. Mendel's laws imposed limitations on the possibility of improving stocks by selective breeding; Heisenberg's uncertainty principle imposed limitations on the predictability of atomic events. Such limitations are contrary to dialectical materialism; hence Mendel and Heisenberg are anathema. Also, Mendel and Heisenberg are easy targets for political attack, Mendel for being a priest, and Heisenberg for being head of Hitler's atomic energy project.

In the Russian translation of Dirac's book *Quantum Mechanics*, published as long ago as 1935, there is a preface inserted which says, "Although this book contains numerous errors and fallacies which are in contradiction with the well-known principles of dialectical materialism, nevertheless it contains so much that will be of value to the judicious student that the editors have felt themselves justified in publishing it without correction or alteration." A very fine piece of diplomacy on the part of the editors. With the help of a few sops like this to the rabid Marxists, the physicists up till 1948 have continued to work and publish results, making full use of Heisenberg, Dirac, or anybody else, and their papers are as sensible as anybody else's.

Now just as in biology a Lysenko can get good results in practice, whether his theory is sound or not, so in physics it happens that nuclear fission is one phenomenon that can be fully understood and exploited, using only the crudest kind of quantum theory. In fact, fission is almost a "classical" phenomenon, in the sense that it takes less quantum theory to understand it than it does to understand the reaction hydrogen plus oxygen goes to water. Not only fission, but most of nuclear physics as it has so far developed, is likewise independent of highbrow theory. In view of this, it would not be at all surprising if the Russian government, wishing to get ahead as fast as possible with atomic energy, should support physicists who work on practical developments in empirical nuclear physics, and should attack those who work on the deeper theories, which everyone agrees will not be of much help to the experimenters in the foreseeable future. It is greatly to the credit of Soviet physics that so far no Lysenko has arisen among the physicists to lead the first group in attacking the second. Perhaps also the government's tolerance of the theoreticians is an indication that the Soviet atomic energy project is going ahead smoothly without them. But this is wild speculation.

Trofim Lysenko was a plant breeder who did tremendous damage to biology in the Soviet Union, using the power of the Communist Party to kill or imprison many leading biologists. He was a fanatical Marxist and believed that modern molecular biology must be suppressed because it was incompatible with orthodox Marxist dogma. As a result of Lysenko's persecutions, a whole generation of Soviet biology was devastated. Fortunately, the physicists were able to stand firm against Marxist dogma, and no Lysenko arose to devastate Soviet physics.

On Tuesday I went to have a quiet talk with Oppenheimer reviewing my seminar talks. He was extremely pleasant and said he agreed with all my main contentions. He had no concrete proposals to make for going further with the theory but advised me to follow my destiny and go on thinking about it until I had squeezed all I could out of it. At the end he asked me what I was intending to do after this year, and when I said I would go back to England, he warmly approved, saying I should try hard to resist the temptation to settle permanently in the States. At the end he said, "You know Dirac and Bohr feel that their proper place is in England and Denmark, and I have an arrangement that they can come to the institute one year in three or thereabouts, so as to keep in touch with people over here. Certainly we shall be able to do something of the kind for you too."

This kind of talk is vastly satisfying to my ego. But I think it is rather silly. Oppenheimer has absolutely no evidence on which he can place me in the same class with Dirac and Bohr, and it is far from clear to me whether I shall ever achieve their kind of distinction. And in any case it would do England no harm if I stayed here for a few years to learn some physics before going back to teach it there. I think his remarks are chiefly interesting as a key to his own character and an explanation of his previous behaviour. It is just this sudden and exaggerated enthusiasm which he showed when Schwinger first produced his theory, and the sudden and exaggerated lack of enthusiasm with which he viewed Schwinger and Feynman when I began my talks. He is a curious mixture, so cool and accurate in his speech and appearance and so nervous and unstable inside.

Sheila Power and Ning Hu arranged on Wednesday to make a joint expedition with me to the Princeton Art Museum where there was a special exhibition of Chinese painting. It was Ning Hu's last day in Princeton. We had with us also Professor Liu, a delightful man, old and round and exactly like the philosophers depicted in some of the paintings; he is a great scholar (in Chinese classics) and is professor of Chinese here. He was for some years a lecturer at Oxford. The collection was extremely good, just three rooms and as much as one could digest in a morning. Professor Liu told us the personal histories of the painters, the things to look for in the paintings, and translated for us the poems which most of the paintings had written beneath or beside them. It was a delightful morning, and I have never before understood and enjoyed Chinese painting so well. Ning Hu maintains that there are only two outstanding artistic achievements, and he places these two on an equal footing. They are, Western music and Eastern painting. You will realise at once what a difference it makes, when you are confronted with twenty paintings from every school and region of China and from eight centuries of time, to be told which is which and why it is different from the others. Just as it is easier to enjoy European painting after you have been told, and grown accustomed to, the difference in time and place between Botticelli and Constable.

DECEMBER 11, 1948

Bethe was here again on Wednesday, with great news from Cornell. The three places where big synchrotrons are in an advanced stage of construction are Berkeley, Massachusetts Institute of Technology, and Cornell. Of these places, Berkeley makes the most noise and spends the most money, MIT next, and Cornell least. All three machines are now completely built, and it only remains for difficult final adjustments to be made before they will work. A week ago, as a result of some clever tricks on the part of Wilson, who is in charge at Cornell, the Cornell machine produced a beam of electrons. Wilson had the pleasure of ringing up Edwin McMillan (the original inventor of the synchrotron) on the long-distance phone to California, and telling him how to make his machine work. Three days later, the Berkeley machine produced a beam. The beams so far are at low energies. But once you have a beam it is not difficult to increase the energy gradually to its full value, and this should not take more than a few weeks. The bad situation is when you don't have a beam and don't know why you don't have a beam. When they get the beam up to full energy, then we shall begin to learn something about mesons and nuclear forces. I am planning a long visit to Cornell as soon as this begins, and Bethe said he would be glad to see me.

DECEMBER 24, 1948

There has been great excitement over a new discovery by Powell at Bristol. Powell always keeps one step ahead of everyone else. A year and a half ago he discovered the so-called pi meson and helped very much the people at Berkeley who subsequently made the animals in their machine. Now he has a new meson which is called a tau meson, three times as heavy as the pi. This is likely to be for several years beyond the capacity of any machines to make, because the larger mass takes a proportionately large energy to create it. Powell found his meson in a photographic plate which he put at the top of a mountain in Switzerland. The main reason for the success was that he was using a new and improved type of photographic plate developed by a man called Berriman at the English Kodak company. The whole research is a joint effort between the Bristol group and the photographic companies.

8

WELL, DOC, YOU'RE IN

THE SPRING OF 1949 was the high point of my life as a scientist, when I enjoyed for a few dizzy weeks the status of a rock star. I began the year by writing a second long paper, forging out of quantum electrodynamics a practical tool for the accurate calculation of physical processes outside the nucleus. I was confident, with a confidence shared by Oppenheimer, that the new tool could be extended to give us an accurate theory of nuclear processes. I looked forward to a second triumph, when the mysteries of the nucleus would be explained. After 1949 my hopes gradually faded. Accurate high-energy experiments showed that nuclear processes are far more complicated than I had imagined. Fifteen years later Murray Gell-Mann and George Zweig discovered quarks, which gave us a deeper understanding of all aspects of the subatomic world. It turned out that every nuclear particle is a little bag of quarks. This discovery opened the door to beautiful new theories of nuclear interactions, but made the calculation of the consequences of the theories more messy. Without quarks, my theory could never have succeeded.

JANUARY 22, 1949

Next week we shall all be in New York at the annual Physical Society meeting. Oppenheimer will give a half-hour talk reviewing the state of physics, which he should do very well if he is up to his usual form. He is particularly good at defining problems and issues when a subject is in a confused state. A good example of his style of presentation occurred the other day at lunch, when I unexpectedly asked him whether he would advise me to choose Birmingham, Bristol, or Cambridge as my next home. He replied without a moment's hesitation, "Well, Birmingham has much the best theoretical physicist to work with, Peierls. Bristol has much the best experimental physicist, Powell. Cambridge has some excellent architecture. You can take your choice." I made up my mind for Birmingham and wrote a letter to Peierls accepting his offer of a fellowship.

This week I began once more to think in a fundamental way, having got the second paper off my mind. The second paper was still concerned with physics outside the nucleus. Now, having got the theory into such a handy and powerful shape, I have decided to sally forth and apply these methods to the nucleus. Almost at once I began to get encouraging results.

Today Bethe came over from Columbia. In the afternoon we had a conference with various people reporting on what they had been doing. Several people of the younger group talked about their calculations, many of which have been done with Feynman methods and have given sensible results. At the end I got up and told them one of the things I had found in the last three days since I started thinking about the nucleus. I think it really startled them, because one of the oldest and most intractable difficulties of the nuclear theory dissolved under their eyes. Always in the past there has been an unbridgeable gap between two kinds of nuclear theory. There is the so-called phenomenological theory, in which the particles are treated as having a certain size, and their properties can then be described with some degree of resemblance to the facts, although it is known that this theory cannot be right since it does not agree with relativity. The other theory is the field theory, which is the only theory which is properly relativistic, and this theory treats the particles as points without size and always gives forces between the particles which are completely wrong and often infinite. What I did in my talk was to take a simple example of the field theory and treat it with the new methods. In no time at all one could see that the point particles were behaving as if they had a finite size, which is just what one wants them to do, and that the forces between them were sensible. Oppenheimer was extremely pleased with this and produced a paper which he had written in 1941, in which he had written down exactly the same formulae as I had for the nuclear force. In his paper these formulae were guesses, based on physical intuition; and now they can at least roughly be derived from a consistent theory. The whole thing is still in a preliminary stage and may not turn out as well as it promises. If it goes, it will give me plenty to think about and enough material for another big paper before long.

Roughly speaking, the difference between my view of the world and Oppenheimer's has been narrowed down to a question of how much can be got out of the nucleus. We are agreed that the existing methods of field theory are not satisfactory and must ultimately be scrapped in favour of a theory which is physically more intelligible and less arbitrary. We also agree that the final theory should explain why there are the various types of particle which we see and no more. However, Oppenheimer believes that the nature of the nuclear forces will itself give us enough information on which to build the new theory. In other words, the nuclear forces will not be describable at all except in terms of a theory which explains the existence of elementary particles. I take a more pessimistic view, that we shall be able to give a complete account of the nucleus on the basis of the present field theory. If I am right, the discovery of a finally satisfactory theory of elementary particles will be a much deeper problem than those we are tackling at present and may very well not be achieved within the framework of microscopic physics. Probably the reason for our disagreement is largely a matter of history. Oppenheimer has spent all his life seeing the field theory fail on one problem after another, whereas I have grown up during one of its brief periods of success.

Bethe told me he was very glad I had decided to go to Birmingham. He had a rather despondent letter from Peierls, saying that he had nobody at Birmingham who knew anything and was himself getting too old to supply all the imagination for the department. From this it looks as if I shall have it pretty much all to myself, especially as Peierls spends a lot of his time on official business. If it is like this, I shall have a tough time trying to carry on my own research and also teach the other people physics. However, it is amazing how one can adjust oneself to circumstances.

Oppenheimer told me that when he came back from Europe as a young man about my age, he had the same kind of problem, being offered jobs as professor at Berkeley and also at Pasadena four hundred miles south, two places which were at the time equally lacking in people who knew any theoretical physics. For some time he hesitated, then accepted both jobs and divided his time between them. And he very soon had under him the best school of theoretical physics in the country.

His students mostly migrated with him north and south each year, spending the fall in Berkeley and the spring at Caltech in Pasadena.

JANUARY 30, 1949

I talk some more about the history of Oppenheimer's relation with the government. At the end of the war, the contact between the government and the scientific people was practically nonexistent; there were a lot of scientists who were clamouring loudly for internationalization of atomic energy, without being very specific about the methods for achieving it; and the government on the whole was not inclined to take them seriously. In this atmosphere, it was largely the personal persistence of Oppenheimer, who walked uninvited into the State Department and pleaded with the officials to do something about it, that resulted in the government accepting the idea of international control. He succeeded in winning and maintaining the confidence of these officials, and avoided on the one hand giving the appearance of being a fanatic and on the other hand accepting any essential compromise on the main issues. It is unfortunately only too easy to think of many scientists who in his position would have made either the one or the other mistake.

The negotiations aimed at an international control of nuclear energy and nuclear weapons continued for several years at the United Nations. There was probably never any chance that an effective international control would have been acceptable to the Soviet Union. The most that Oppenheimer could achieve was a set of proposals for international control, agreed by both parties in the U.S. government and sincerely offered to the international community. The proposals were rejected by the Soviet Union. They were a serious effort by the United States to change the course of history and put a stop to the nuclear arms race.

The New York meeting was from my point of view a fantastic affair. All my friends from Cornell, Ann Arbor, and Princeton were there, besides many others, and it was a continual social gathering from morning till night. On the first day the real fun began. I was sitting in the middle of the hall and in the front, with Feynman beside me, and there rose to the platform to speak a young man from Columbia whom I know dimly. The young man had done some calculations using methods of Feynman and me, and he did not confine himself to stating this fact but referred again and again to "the beautiful theory of Feynman-Dyson" in gushing tones. After he said this the first time, Feynman turned to me and remarked in a loud voice, "Well, Doc, you're in." Then as the young man went on, Feynman continued to make irreverent comments, much to the entertainment of the audience near him. Later on Feynman himself spoke on his own work and created so much uproar with his clowning that the audience voted him twice the usual time for his talk.

On the second day Oppenheimer gave a presidential address in the big hall, and such is the glamour of his name (after being on the cover of *Time*) that the hall was packed with two thousand people half an hour before he was due to start. He spoke on the title "Fields and Quanta" and gave a good historical summary of the vicissitudes of field theory. What was overwhelming was that at the end he spoke in enthusiastic terms of the work I have been doing and said that it was pointing the way for the immediate future, even if it did not seem deep enough to carry us farther than that. I just sat there feeling small. You can just imagine what my life was like for the next twenty-four hours; one person after another pursuing me and asking to be told all about it, and sometimes several patiently waiting their turn. I am becoming a Big Shot with a vengeance.

I don't know why Oppenheimer should have been so indiscreet. Perhaps it is partly because I am English, and he is much concerned with impressing upon the nationalistic war generation the fact that science is an international activity. However that may be, I believe I am wise enough to enjoy this sort of success without being taken in by it; if I were not, I have the example of Feynman to instruct me.

I had sent to my parents a copy of "The Open Mind," an article by Oppenheimer that appeared in the Atlantic Monthly, *to give them a taste of his style and personality. They found the article puzzling, and I wrote this letter to explain why.*

FEBRUARY 15, 1949

Oppenheimer is always so successful in avoiding oversimplifying a problem, and touches so delicately the state of mind of a man who is having to move between the world of ideals and the world of realities, that when he has finished, you ask yourself, "Now just what did he say?" And the answer is that he uses words deliberately to suggest rather than to define, and means to say different things to different people. But I think that to nobody were these remarks meant as an appeal for the abolition of secrecy and coercion. To my mind they were first of all an appeal for patience, and balance, and humility before the unknowable forces of history. They were addressed to the college president from the prairies, who wanted his students to cure the world's ills by their own efforts. They were also addressed to the scientists who think they could solve the problems of politics if it were not for the politicians. But in addition they were addressed to the politicians, and to the ordinary public which follows the lead of the politicians; to point out to them the essential place in scientific ethics of the ideals of openness and noncoercion, and the vast practical difference between making a compromise between honestly held ideals and expediency and not having any ideals. I don't know how you can read through this article without being struck by the grimness of Oppenheimer's philosophy, and his deep awareness of the irresoluble conflicts and tragedies of the present age. It is not for nothing that his chosen model is Lincoln, a liberal not at all of the prosperous aristocratic kind, and a man whose life was spent in the exercise of coercion.

If I know Oppenheimer at all, he had thought all these things out thoroughly before he started making atomic bombs. It was Lincoln's great achievement, in the long run, that he was able to make effective use of coercion and still create a legend of honesty and magnanimity which influenced his contemporaries and his remote successors. I believe this is not a bad solution. It was always Oppenheimer's aim, and one which at least partially was achieved, that the American government and people should be honestly convinced of the desirability of handing over their atomic weapons to an international authority. Whether such an international authority could exist, now or in the foreseeable future, was a separate problem and one which only history could answer. But he felt that by making a sincere and extended stand for international control, the American government and he himself could to some extent expiate the guilt of Hiroshima and could create a tradition which would endure like Lincoln's to influence the minds of future generations.

FEBRUARY 28, 1949, CHICAGO

On Thursday we had Feynman down to Princeton, and he stayed till I left on Sunday. He gave in three days about eight hours of seminars, besides long private discussions. This was a magnificent effort, and I believe all the people at the institute began to understand what he is doing. I at least learnt a great deal. He was as usual in an enthusiastic mood, waving his arms about a lot and making everybody laugh. Even Oppenheimer began to get the spirit of the thing and said some things less sceptical than is his habit. Feynman was obviously anxious to talk and would have gone on quite indefinitely if he had been allowed; he must have been suffering from the same bottled-up feeling that I had when I was full of ideas last autumn. The trouble with him is that he never will publish what he does; I sometimes feel guilty for having cut in front of him with his own ideas. However, he is now at last writing up two big papers, which will display his genius to the world; and it is possible that I have helped to make him do this by making him conscious of being cut in on, which if it be true is a valuable service on my part.

The day before Feynman came, Eugenio Lattes was in Princeton. He gave a first-rate talk to our seminar about the work which is being done at Berkeley. Afterwards I had the luck to talk with him for a long time in private and hear some of his story. It was good for me, for reasons which will soon be clear. Lattes is the young Brazilian physicist, now aged twenty-four, who is the co-discoverer of the heavy meson. He started his meteoric career by joining Powell's team at Bristol, and with Powell he did very well, and his name was with the others on the paper announcing the finding of the meson. Then last spring he went over to Berkeley, and as soon as he got there, he became the discoverer of the first artificial meson, by the simple process of noticing that the Berkeley people had been developing their photographic plates for too short a time. When he developed some plates the proper way, there were the mesons. Now it happened that the newspapers and the government in Brazil heard about these events and were not well-enough informed to know what it was really about. And so when Lattes went home to see his family last summer, he found that five radio networks had come with microphones to the airfield to broadcast his homecoming speech, and he was received in state by the president of the republic and the governors of three states. In fact, he was the greatest physicist who ever lived.

Now comes the remarkable part of the story. Lattes, in spite of having been the victim of this fantastic accident, is a good physicist. He is also a modest and far-sighted young man. And so he decided to push for all he was worth, while the going was good, for the endowment of a decent Institute of Physics, with enough money to pay for students to study there and to travel to the United States and Europe, and for a good permanent staff of active physicists. Thus he plunged headlong into the arena of politics and high finance and went around interviewing all kinds of people, mainly millionaires and politicians, and in a few weeks he had enough money to get the thing launched. And it is really going to be built, in São Paulo, where Lattes himself was a student.

Lattes himself takes a very long view of all this. He says, with a country as big and as backward as Brazil, it is not to be expected that a first-class centre of pure science will spring up. Rather he hopes that the opposite will happen, that his institute will be a stimulus to the development of the country and will be exploited by the industries from a practical point of view, as a place where engineers can get a first-class scientific education. He says the most important immediate objective is merely to raise the prestige of science, so that the few people who do get a higher education do not consider science beneath their dignity. It made me feel very strange, to be talking to this young man, a mere boy in appearance, who has had all these responsibilities dropped into his lap by the ironic Fates. Amongst other things, by going back to Brazil and devoting himself to these projects, he is deliberately sacrificing what would be a fine opportunity to continue his own work at Berkeley. He says he will have plenty of time. But most striking of all is the objectivity and sense of humour with which he regards all these events. What a lucky thing it is that it was Lattes this happened to and not someone else! It seems my own history at the moment has a little (very little in comparison) of the same quality, so I am grateful to Lattes for setting such an admirable example.

Shortly after our conversation, Lattes returned to Brazil and successfully raised money to support programs of physical research. After a promising start, he became entangled in financial and political scandals. He escaped from the turmoil but abandoned his hopes of leading a scientific revolution in his homeland. He spent the rest of his life as a professor of physics at various universities in Brazil.

MARCH 11, 1949

I must now start to tell about my adventures in Chicago. This will be a long story, and even so I shall not be able to do the real thing justice; never have I met so many outstanding people and talked so much and eaten so much and drunk so much and slept so little in one week. The head of the cosmic ray division, Professor Marcel Schein, was responsible for my welfare. Schein is a dear old boy, rather garrulous but still energetic, who came to Chicago from Zürich many years ago. When I arrived he was very excited because he had just got his first batch of electron-sensitive photographic plates from the Eastman Kodak company; these plates are a great advance on anything previously known, but are still very hard to make and to handle. The first such plates were made by the English Kodak people at Harrow, but they are so temperamental that they do not usually survive being flown across the Atlantic; that is why Powell at Bristol has been enjoying a monopoly of them for some months. It was important to send such good plates as these up to one hundred thousand feet to record what is happening at the top of the atmosphere, and Schein had the necessary apparatus for this. In England there is not much use in sending up things to one hundred thousand feet, because they will usually come down in the sea. So there were about twenty people gathered on Thursday morning to watch the historic occasion when Schein sent up his first batch of plates. It was a filthy morning, with clouds and strong wind, but they had no choice but to go ahead with the launching, because the plates deteriorate so rapidly. The launching was done with great efficiency, with ten light rubber balloons about ten feet across; each balloon was held by one man on a string, and the string was then fastened to the little box containing the plates and a recording altimeter. One by one the men let go of their balloons, until finally the last man let go and the whole apparatus shot up into the air and majestically disappeared into the clouds.

Two days later the plates returned in triumph, having been brought home from somewhere in Ohio after a successful flight of about ten hours at one hundred thousand feet. And on Monday the plates were developed, and I was looking at them through the microscope. When you look at the plates like that, you see nothing but a maze of rather uninteresting tracks, made by fast particles of unknown speed and nature. It takes several weeks before you have scanned the plates to make sure there is nothing spectacular recorded. To sort out all the detailed information in such plates is a formidable job, which I doubt Schein's group will ever do adequately. Schein is not enough of a slave-driver; also, he lacks other qualities which make Powell unique in this field.

For the physics talk, I had the satisfaction of being moved from a smaller classroom into a larger one because too many people turned up. I spoke mainly about the way the Schwinger theory and mine link up with each other, and I was very polite to Schwinger. At the end of it, Edward Teller, of whom I shall have more to say later, asked the following question: "What would you think of a man who cried, 'There is no God but Allah, and Mohammed is his prophet,' and then at once drank down a great tankard of wine?" He said that he himself would consider the man a very sensible fellow. Teller is a Hungarian, being one of the famous four, the others being Wigner, Szilard, and von Neumann, who grew up around the same time in one particular district in Budapest, and who are now all in outstanding positions in the United States. Teller made the move from Europe to America in 1935.

Teller to me has always been something of an enigma; he has done all kinds of interesting things in physics but never the same thing for long, and he seems to do physics for fun rather than for glory. However, during the last few years there have been reports that he has been engaged in perfecting the most fiendish engines of destruction; and I have always wondered how such a man could do such things. In Chicago I found without difficulty the answer; I started a long argument with him about political questions, and it appears that he is an ardent supporter of the World Government movement, an organization which preaches salvation in the form of a world government, to be set up in the near future with or without Russia, and to have sovereign powers over the economic and social policies of its member nations. Teller evidently finds this faith soothing to his conscience; he preaches it with great charm and intelligence; all the same, I feel that he is a good example of the saying that no man is so dangerous as an idealist.

Of the other people whom I met, one was Fermi, the great Italian physicist and one of few men alive who has done first-rate work both in theory and in experiment; during the war he was the boss of the Chicago Metallurgical Laboratory which built the first pile and ran the plutonium production project. He came up a few weeks ago with a brand-new theory concerning the origin of cosmic rays. He listened with close attention to everything I said, and said not much himself; in this he differs from Oppenheimer. The most striking thing about these people, and also their wives whom I met as I went from house to house and from family to family, is how happy they seem to be; all of them say they have never found any place on earth so pleasant to be in as Chicago. There seems to be an exceptionally free and easy atmosphere, rather like Cornell, and with the added advantages of a metropolitan city.

Another remarkable man I met was André Weil. I was lucky enough to go for a long walk with him along the lakeshore and to discuss the world's problems. He is one of the finest living mathematicians, still quite young (about thirty-five), and with a stormy career behind him. He is a Frenchman but has always regarded himself as a citizen of the world; as soon as he had taken his degree, he went to India for two years and learned Hindustani and Indian literature while at the same time teaching mathematics and trying (he says unsuccessfully) to inspire the Indians with a little enthusiasm for the subject. After this he returned to France for a time but was then overtaken by the war; being a conscientious objector, he was put in prison, and there wrote an epoch-making work on topological groups. In the French defeat and ensuing confusion, he managed to escape by a devious route to Brazil and was a professor of mathematics at São Paulo for three years, where he achieved, he said, slightly but not much more than at the Moslem University in India. Then finally he found a haven of refuge at Chicago, where he has lived happily ever after. He takes an extremely dim view of the prospects of the world and expresses this view with the best epigrammatic polish. Particularly he was horrified by the moral decay which he believes to have overtaken France, and he says it is graft and dishonesty which keep Brazil backward much more than material poverty. I was enthralled with his conversation, not so much for the generalizations as for the particular stories about India, Brazil, etc.

André Weil was the brother of the famous writer Simone Weil who wrote about religious and mystical experiences. Simone died young. André had enormous respect for her and continued for all his life to mourn her death.

Apart from all these great men, there were innumerable other people who came to one or other of the parties with which I was entertained. I felt at the end of a week that I knew Chicago and its people much better than I ever knew Cambridge, and it is precisely this open friendliness that everyone likes about it. Several times it was hinted to me that I could do worse than to settle down in Chicago, and I felt very much inclined to agree.

My position at Birmingham is more or less fixed. Peierls says he can get me a fellowship with no duties to speak of, and I have officially resigned my Trinity fellowship. The details of the Birmingham fellowship will be fixed up in the next month or two. What is more important, Peierls says he can take me in as a lodger in his own house. To live with the Peierls family would be pleasant in many ways, as they are energetic and interesting people.

As you may have guessed, I do not expect that the barrage of invitations and lectures and trips which I am enjoying here will stop when I return to Europe. I intend to travel around a lot under my own steam and try to help pull European physics together. If I am to do this, it is Bohr most of all whose blessing I shall need. I have never met Bohr, and I have undoubtedly a tremendous lot to learn by going and working in his institute at Copenhagen and listening to his views about all sorts of questions.

Niels Bohr was the leading physicist of Europe, second only to Einstein in the world. In his institute in Copenhagen, Bohr had presided over the development of quantum mechanics in the 1920s and '30s. I never went to work at Copenhagen. I got to know Bohr under totally different circumstances in California ten years later.

MARCH 27, 1949

After a mild winter, spring is coming to Princeton, hot but sunny and pleasant. The squirrels have come out of their holes and are playing in the trees outside my window; the children have come out of doors in summer clothes and are playing underneath the trees. No leaves are yet upon the trees, but the grass is green and the crocuses are out. The children are more brightly coloured than the crocuses.

My last trip lasted ten days, with a week at Ithaca. I always feel attached to Ithaca. Many of my old friends were still there and made me feel nostalgic about the place. I spent a lot of time with Feynman and with Ning Hu. The best thing about the trip was that I spent three days living in the house of the Bethes and so got to know them better than ever before. They are delightful people, especially the younger generation, Henry aged five and Monica aged three, with whom I got on very well. Usually I was given the pleasant job of distracting the children's attention while Mr. and Mrs. Bethe prepared the meals, a division of labour which suited me very well. Henry and Monica are both bright and are already having trouble with the rigid American educational system which discourages strongly any child getting ahead of its age. So strong in fact is Henry's thirst for knowledge that he insists in going to Sunday school every Sunday, although his parents do not like the school and do their best to stop him. I think this says a good deal for the Sunday school. The church is making good use of the opportunities offered to it by the slowness of the educational system.

Mrs. Bethe is thirty-two, a tall good-looking young woman and a very efficient person. When they were at Los Alamos, it is said that she contributed more to the success of the project than he, since she was in charge of the living arrangements while he was only doing physics. They are both German, but they came over to the States independently, and they met first at Cornell where Rose was a student while Hans was already a professor. He is about ten years older. Since both of them are such positive and outspoken characters, it is surprising that they live together as harmoniously as they do and not surprising that their children are argumentative. Sample conversation at breakfast: Rose: "Eat up your cereal, Henry." Henry: "Mummy, are you the boss in this house?" Rose: "Shut up." Henry: "Well, that's not a very nice thing to say, is it?"

I have a sad letter from my German friend Hilde Jacobs, who is now in hospital with (she says) pneumonia. I hope this does not mean

TB, which is fearfully prevalent in that part of the world. I wish I could do something for her. It seems to grow harder and harder to think across the gulf between my prosperity and her misery, and so our correspondence languishes. When she went to stay in Switzerland, she said she asked herself always, *War Gott vielleicht Selber Schweizer?* [*"Was God himself perhaps Swiss?"*] And so to her the United States is something altogether fabulous and dreamlike.

April 5, 1949, Hart House, University of Toronto

I find the people here most friendly and an interesting crowd; but how different from an American university! Head of the maths department is an old boy called Beatty, humorous and beloved of the students, a very good teacher. Under him are various others, all Cambridge-educated and with the Cambridge easygoing philosophy. The man who looks after me is a man of about forty called Robinson. Robinson is the most active of the lot, does a certain amount of research, and helps to edit a newborn journal, the *Canadian Journal of Mathematics*. He is also a good amateur photographer, and his wife is a leading light in the Canadian YWCA. Youngest member of the staff is a Cambridge contemporary of mine called Tutte, a shy and silent young man but apparently settling down happily here.

Bill Tutte had spent the war years at Bletchley as a code-breaker. The British policy of total secrecy prevented his achievements at Bletchley from becoming known until many years later. Together with Alan Turing, he had played a major part in the decipherment of high-level German communications. He remained shy and silent to the end of his life and never received the recognition that he deserved.

All in all, it is a well-organised, respectable, and happy society such as one might find at a good English provincial university (though this university is much bigger, with seventeen thousand students). Almost completely lacking are the crowds of young people, interested primarily in research and not in teaching, and the bustling competitive atmosphere of the American university. Beatty himself is conscious of this lack and worried about it. He says all the bright young people he unearths among his students run off to the United States at the earliest opportunity, and he doesn't blame them. He has neither enough money nor enough glamour to hold them. The great problem is how to provide positions for young people at good salaries in the science faculties without making the other faculties jealous. It is a sad story, and one that is to be found almost all over the world.

Among the physicists here, the chief theoretician is Leopold Infeld. He also suffers from lack of good young assistants; however, he is by nature a rather isolated figure and is quite happy. He works on general relativity and has collaborated a good deal with Einstein. I would like to see more of Infeld. He is, apart from his physics, a highly colourful character, in marked contrast to the British solidity of his colleagues. He is a Polish Jew, a magnificent storyteller and conversationalist, and a gifted writer of English. He has had endless adventures in all parts of the world, knows everybody and all their private scandals, and his tour de force is a scientific autobiography which he published in 1941 at the age of forty-three, and which became a bestseller. It is called *Quest*. In contains, among other things, a delightful and appreciative description of life (prewar) in Cambridge and a sulphurous and malicious description of life (prewar) at the Institute for Advanced Study; and various libellous anecdotes about Dirac and other geniuses. He was glad to make my acquaintance and listen to all the stories I could tell him about Schwinger and Oppenheimer and such people. I expect to see these stories reappear with embellishments in his next book.

9

THE PHYSICIST IN LOVE

APRIL 7, 1949, PRINCETON

The New Yorker has published the description of life at the institute which I enclose as I think you will like it. It is a great deal more objective than the accounts in *Life* and *Time*, though it makes the same mistake of overestimating the importance of the institute to the intellectual life of the country. Also some of the details are subject to a little poetic exaggeration; the girl who looked about eighteen and is an expert on nonassociative algebras is actually Verena Haefeli, aged twenty-six and mother of a fat two-and-a-half-year-old daughter called Katrina.

I am beginning to find, as usually happens to me when I am a few weeks from leaving a place for good, that there are some people here with whom I might have formed some real friendships if I had got to know them a bit earlier. In particular, the Saturday night dances at the institute have been a tremendous help in getting to know people; these began originally just about the time I was starting on my travels to Chicago and elsewhere, so that I have only taken part in them three or four times. The life of the Housing Project, where all the young married people with their families live, has seemed to me until very recently a Garden of Eden from which I was excluded by an unkind fate; it is just in the last week that I have begun to penetrate into this society.

The dances at the institute were square dances organized by John Tukey, then a young professor at Princeton University, later famous as one of the founders of computer science. He was an expert square dancer and knew how to call the dances, so that an international and multilingual group of institute members could catch the rhythm and keep in step.

This afternoon I had a wonderful piece of luck. I was strolling around by myself, taking the air and feeling somewhat melancholy, and I happened to pass Verena Haefeli's house on the Housing Project, and little Katrina was standing at the window and began shouting at me as I went by. This was odd, because I had hardly spoken to her before today. I took advantage of the occasion to go in and see Verena's house and spent a delightful afternoon learning her life history and playing with Katrina. Tomorrow I am invited to join the two of them and two other people in an expedition to the ocean, where there will be picnic and bathing.

Verena was a Swiss mathematician, educated at the University of Zürich and expert in mathematical logic. She had been married to Hans Georg Haefeli, also a mathematician, who had moved to Boston College in Massachusetts after their separation. She wrote an account of her life as a chapter in the book Kreiseliana *(1996), celebrating the seventieth birthday of Georg Kreisel, the famous mathematical logician whom I had known as a student in Cambridge.*

Verena has a tough time of it, being a mathematician and also bringing up her daughter single-handed. She is from Zürich, and her father was an important person in the Swiss Red Cross; both her parents are now dead, and she has knocked about the world a good deal. Katrina is a beautiful child, with golden hair and dark brown eyes, and already has a mind of her own. She has asserted her independence by forming a close friendship with Ocky, the six-year old son of Professor Uhlenbeck, and showing a certain scorn for her own contemporaries. Perhaps Ocky is to some extent a substitute for a father.

George Uhlenbeck was a famous Dutch physicist who had moved to America before the war and become head of the physics department at the University of Michigan. He organized the summer school in Ann Arbor where I got to know him in 1948, and then spent a sabbatical year at the Institute for Advanced Study, where his wife Else became a friend of Verena.

Verena's father Charles Huber spent World War II as delegate for the International Red Cross inspecting prisoner-of-war camps on both sides as allowed by the Geneva Convention. He was mostly visiting Allied prisoners in Germany and German and Italian prisoners in India and in the United States. Soon after the war ended, he was driving his car at night in Germany and died in a collision with a truck that was standing on the road without lights.

Three famous conferences were organized by Oppenheimer to bring together the experimenters and theorists who were leading the revolution in quantum electrodynamics. The first was at Shelter Island in 1947, the second at Pocono Mountain in 1948, and the third at Oldstone in 1949. Bethe and Feynman were at all three meetings and told me what had been done and said. I was invited only to the third.

EASTER SUNDAY, APRIL 17, 1949, PRINCETON

On Sunday we started off for the Oldstone Conference, where the big shots of physics were assembled. Oldstone is a country hotel, about fifty miles north of New York, in a splendid situation overlooking the Hudson, with hills behind it and hills facing it across the river. We had lovely weather for the conference and could sit outside whenever we were not conferring. Since the conference was run by Oppenheimer, that was not often. One of the things which amaze me about Oppenheimer is his mental and physical indefatigability; this must have had a lot to do with his performance during the war. There was no fixed program for the conference, and we talked as much or as little as we liked; nevertheless Oppenheimer had us in there every day from ten a.m. till seven p.m. with only short breaks, and on the first day also after supper from eight till ten, this night session being dropped on the second day after a general rebellion. All through these sessions Oppenheimer was wide awake, listening to everything that was said and obviously absorbing it. For my part, knowing my limited capacity for attention, I chose a comfortable chair and allowed nature to take its course; three times during the conference I slept soundly and unnoticed, and the rest of the time I was the better for it.

On the second day I stood up and talked for about an hour. I found this easy, as I did not have to prepare anything but only to answer questions and summarise what the other young people at the institute have been doing. Schwinger and Feynman were there, also Bethe and Teller and Serber and Rabi, among the people you have heard of, and from Princeton Wigner and von Neumann. Among these great names there were just two small ones, the babies of the conference, myself and Aage Bohr. Aage is a nice fellow, and I was glad to have him there.

Aage was a son of the great physicist Niels Bohr. Aage lived for many years in his father's shadow but later emerged as a leader of European physics, winning a Nobel Prize for Physics in his own right in 1975.

As was expected, the conference did not produce anything startlingly new. The most valuable part of it was a lot of detailed and firsthand reports of the situation in the various big experimental labs, Berkeley and Columbia and MIT. To me, the great thing was to get to know the people and to see how their minds work. I saw a lot of Rabi and of von Neumann, both geniuses of the first order. In the formal sessions the theorists contributed not much, except for some long and confused arguments which were understood by nobody except the arguers and Oppenheimer.

On the way home from the conference, we stopped and looked at the great new cyclotron at Nevis, a village halfway between Oldstone and New York. This is a magnificent machine, belonging to Columbia. It will be running now in a few months. It is rather like the Berkeley machine, only more powerful, and it is more impressive to look at because it has not yet the concrete shield around it. The most remarkable thing about it is the way Rabi has organized it. Everywhere else where a laboratory has undertaken to build a great machine, almost everybody became involved, and gradually the rest of the work of the laboratory came to a standstill. Rabi swore that this would not happen in his lab. With iron determination, he decreed that the whole construction job should be done by two physicists, using otherwise only engineers and labourers. That is the kind of man Rabi is.

Every year the American Physical Society holds one big meeting in Washington and smaller meetings in other places. To be invited to give a plenary talk at the Washington meeting was a big honor.

MAY 1, 1949, PRINCETON

The Washington meeting lasted three days and was on the whole very successful. Fortunately my talk was on the first day, and once that was over, I could devote myself to meeting my innumerable friends who all seemed to be there. Being in the most fashionable branch of physics, I was put into the largest auditorium, a grandiose monstrosity with enormous gold-painted columns stretching up to a domed and bright blue roof. There was an efficient amplifier system so I did not have to shout. About half an hour before I was due to start, I came in and had a look around this place, and the sight of it made me so nervous that none of my previous agony at Chicago and elsewhere could faintly compare with it. For that last half-hour I was in a terrible state, sitting in a chair and sweating all over and feeling I could not even stand up. I was seriously considering sending a message to the chairman of the meeting that I had been taken ill and would not be able to talk. I do not know why it happened like this; perhaps I was worn down by the cumulative effect of the last two months; I had originally intended to write out my talk in full, but from laziness and lack of time had not done this and was equipped only with my usual rough notes. I somehow lived through that half-hour and heard the clock strike the time I was due to start. I staggered into the auditorium and up to the platform while Rabi announced my name and history. As he finished, there was a hushed silence, and I felt ready to collapse completely, but mysteriously the instinct of showmanship triumphed over the instinct of terror. Without thinking, I jumped up onto the platform instead of walking round by the steps. Once I was up on the platform, I began to talk, and when next I paused to consider my situation, I was halfway through my speech and feeling fine. I went a little too fast and finished with some minutes of my time to go, but as it was half-past five and the end of a long day, this was a fault in the right direction. After the talk, Oppy came up and shook me by the hand and said I had done very well, and so did various other people whose praise is more to be valued because it was spontaneous.

MAY 15, 1949, PRINCETON

I write at the end of a long and lovely day, spent with my two newest and firmest friends, Verena Haefeli and her daughter Katrin. We started out, the three of us, at nine o'clock in the morning in Verena's car and drove sixty miles to a town called Lakewood. The weather is now cooler, breezy and sunny and ideal for driving. The New Jersey countryside to the east of us is flat and fertile, and the wheat and barley are already almost full grown and only waiting to ripen for the harvest; with these wide open spaces and the richness of the land, to drive along the open road is an exhilarating experience, giving one a vivid feeling of freedom and of the bounty of nature. Arrived at Lakewood at ten-thirty, we descended to a more prosaic plane and went in search of the police station. Here Verena had to appear to answer a summons, which she had been given a week earlier for failing to stop at a stop sign nearby when driving through. We found the chief of police, who was not a formidable person, and he dealt with the whole case in a few minutes, ending up with a fine of ten dollars for Verena. This being duly paid, we set off once more to the east. We came soon to the sea and drove for some way along the coast, which is here low and sandy, with some breakers and a good salt taste in the air. We stopped at a fishing village and bought three fresh flounders for supper. Then we went a little inland and found a place for our picnic lunch, a little clearing in a pine wood, with a floor of dry pine needles to sit on, and with pine trees swaying in the wind over our heads. Here we laid out our provisions and started in earnest on cold roast chicken, bread and cheese, tomatoes and oranges. And afterwards we exercised ourselves mildly by holding the four corners of a rug outstretched and bouncing Katrin up and down in the middle.

In the evening I came round to Verena's house an hour before supper to wrestle with the task of dissecting three fresh flounders into fillets. This task was accomplished, not without some confusion and a good deal of ribaldry; we had some difficulty persuading Katrin that the fish did not really mind being beheaded and made into fillets. After all this work we sat down to a substantial supper.

Last Tuesday I went up to New York to talk to the seminar at Columbia and to see Rabi's experiments. I arrived in time for lunch, and then spent a strenuous afternoon touring the laboratories. Rabi himself escorted me some of the time, and I learned a great deal about the way these things are done. What makes all this work so impressive is that it is all done in ten little rooms on a single floor of the Pupin building. Not one experiment takes up more space than ten feet square, and Rabi has no desire to expand his activities at the expense of the people who work in other branches of physics on the other ten floors of the building.

Today came a letter from the Physikalische Anstalt der Universität Basel. They are holding an International Physics Conference, half in Switzerland and half in Italy, during the first half of September. Specifically, they will meet in Basel from September 5 to 9, then in

Como from September 11 to 16. I am invited by the organisers to give a talk at the Basel session on the stuff I have been doing. Also talking there will be Rabi, Schwinger, and various other distinguished people. This is a wonderful introduction for me into the European physics world. Also I am pleased with the chance of seeing Switzerland and Italy. My only anxiety is that I have won these laurels a little too cheaply; I had thought I might do some solid work in Europe before this started to happen. I shall soon be in danger of forgetting how to sit down quietly and learn something. However, I am keenly looking forward to the jaunt, and I intend to make the most of it.

Rabi has been travelling around recently and seems to thrive on it, both physically and intellectually. When I was in New York, he told me about his recent trip to Japan. He went to see the emperor and found him a remarkably pleasant and intelligent man; he knew a good deal about Yukawa and other Japanese scientists with whom Rabi has been consulting. When he came home from this trip, Rabi went to see Truman and had a chat with him about the world situation. He found Truman in an extraordinarily optimistic mood. This was so overpowering that he has almost become an optimist himself. I have had letters from Germany, one from my correspondent Richard Käferbock in Stuttgart announcing the birth of a son, and full of enthusiasm. I am pleased with this; it means that new life is at last returning; when I was in Germany two years ago, there were practically no babies.

If life is a drama, here is the beginning of act two. These letters to my parents are not only a record of events but also a piece of the drama. It was important for me, as I was strutting on the stage of my life in America, to have my parents as an audience participating in the drama. Writing the letters was a part of the play.

May 23, 1949, Princeton

I will not make this a long letter, because in these last days my mind has been completely occupied with problems even more incommunicable than those of mathematical physics. In short, I am in love.

And being a person who is unable to concentrate upon more than one thing at a time, I no longer make even a pretence of working or of listening to the people who are still persevering with their calculations at the institute. And this is all to the good, since I needed a holiday anyway. Nor do I find it possible to be interested in politics and to read the newspaper every morning; and this is all to the good too, since my eyes have long been overdue for a rest.

What is not all to the good, and what is a cause of my great woe, is that Fate has played me the trick of setting down before my eyes a woman of altogether exceptional character and quality, with intellect and interests to match my own, and then taking her off to the ends of the earth before I have recovered from the first shock of recognition, and before my slow wits have time to do more than stretch forth my hand in greeting to her. Perhaps this sad little drama is not yet played through to its end. I have still one week left, before this woman, Verena Haefeli, departs for California. So I will write no more about her now; in due course you will hear the end of our story, what little there may be to tell.

May 31, 1949, Princeton

This afternoon at two oclock, Verena and Katrin and Cécile Morette, in Verena's car, set forth on their journey to California. Their car was loaded to the roof with provisions and household equipment, and their departure was attended by a great throng of their friends and admirers. It is a hard and ambitious trip for a pair of girls to undertake, especially with Katrin to be looked after; however, as I know that Verena is very competent but also as tough as a horse, I have not much anxiety on this score. They will take it slowly, with a few days halt in Colorado, and expect to be in San Francisco in two weeks. Then Cécile goes for a few weeks to visit Berkeley, and Verena goes for three months to Stanford, some thirty miles to the south. When Verena went away, I sent after her a letter, saying thank you for all she had done for me. Into it I put a verse from a poem of Day Lewis, out of the

little collection which you gave me for my birthday (1948). The poem
is addressed to Thomas Hardy the novelist; it is not great poetry, but
I felt it expressed to perfection what I wanted to say to Verena. And
since it was your gift to me, I write it down here again for you.

> Great brow, frail frame gone.
> Yet you abide
> In the shadow and sheen,
> All the mellowing traits of a countryside
> That nursed your tragi-comical scene;
> And in us, warmer hearted and brisker-eyed
> Since you have been.

The poem is "Birthday Poem for Thomas Hardy." Cecil Day Lewis was my
favorite poet, speaking for the younger generation in the prewar and wartime years.
He is now less famous than his actor son Daniel, who speaks with the same passion
on stage and screen.

JUNE 6, 1949, PRINCETON

I can now see what I did not realise while Verena was here, that
she occupied a central and important place in this Housing Project
community. Partly because of her peculiar circumstances, partly
because of her natural beauty and attractiveness of character, everybody knew who she was, and most people considered themselves her
friends. During her last days while she was clearing up her house,
there were any number of people in the adjoining houses who would
have been only too glad to come to her assistance, but seeing that I
was in attendance upon her, they tactfully stood aside. I have been
fortunate, now that Verena is gone, in inheriting a large measure of
the interest and goodwill that she left behind her. It is this inheritance
which now makes my days happy and makes me for the first time
think of Princeton as a home I shall be sorry to leave.

JUNE 26, 1949, PRINCETON

Said a final good-bye to Oppy; we had nothing much to say to each other, and I hardly gave another thought to the matter; how different from the violent feelings of mixed respect and impatience with which I met him nine months ago. As you may have heard, there has been a major crisis in recent weeks over the U.S. Atomic Energy Commission and its policies. Oppy has been completely absorbed in this. He is the one scientist to whom the senators and people are prepared to listen with respect, and so he has a tremendous responsibility. He has fought with all his strength for the freedom of science from political pressures, which is the fundamental issue at stake in the present fight over David Lilienthal. As a physicist, Oppy has his limitations, but as a politician he is really outstanding.

In the many volumes that have been written about Oppenheimer, primary emphasis is always given to the security hearings of 1954 which resulted in the loss of his clearance. He becomes the central figure in a tragedy, and the loss of the clearance becomes the central event in his life. He himself did not look at things this way. He always said that the security hearings were not a tragedy but a farce. Much more important in his view were the nine years, from 1945 to 1953, when he enjoyed enormous respect inside the American government as well as in the world outside. During these years he used his unrivaled influence to drive American nuclear and military policies in directions that he considered wise. David Lilienthal was a prominent liberal who had been the head of the Tennessee Valley Authority and was the first chairman of the Atomic Energy Commission. Oppenheimer's support of Lilienthal was an important part of his political agenda.

SUNDAY, JULY 3, 1949, PRINCETON

In one of your letters you thank me, because I write to you openly about Verena and the other important events of my life. I think the reason I write so openly is just this, that all these adventures in this strange new world are still somewhat unreal to me, and in writing to you about them, I bring them in contact with my familiar world and lend them some of your reality.

After this letter there is a gap of a year. I spent that year mostly in Birmingham, making frequent visits to my family in London, so there was no need for letters. In August 1949 I met my pen-friend Hilde Jacobs in Germany and spent a week with her, camping and walking across southern Germany all the way to the Swiss border at Lörrach, close to Basel. After two years of literary exchanges, we got to know each other as real people. We liked each other and agreed to stay friends but not to meet again. I was then deeply involved in my interrupted friendship with Verena, and further meetings with Hilde would bave made a complicated situation even more complicated. Hilde and I stayed in touch by letter for sixty years after that. I sent her family news, and she sent me poetic meditations. She lived for some years in Ireland and later returned to Germany with a job as student adviser at the University of Wuppertal. My last letter to her was returned as undeliverable in 2011. The last I saw of her in the flesh was at our parting at Lörrach in 1949, when I walked over the border into the paradise of Switzerland, and she walked back into the desolation of Germany.

I arrived in Basel in time for the International Physics Conference. I was welcomed by Markus Fierz, professor of physics at the University of Basel and local organizer of the conference. We came together into a room where Wolfgang Pauli, the great physicist from Zürich, was talking in German to a group of respectful listeners. Pauli was famous for his strong opinions and his sharp tongue. Pauli told how Schwinger had come to Zürich and explained the new American physics clearly, not like the nonsense that Dyson had been writing. At that moment Fierz pushed me forward and said, "Professor Pauli, please allow me to introduce you to Professor Dyson." Pauli replied, "Oh, that does not matter, he does not understand German," and continued his discourse. Afterwards Pauli always treated me with great respect, and we became good friends.

I lived for a year as a lodger in the home of Rudolf Peierls in Birmingham, with a lively family of four children brought up by his Russian wife Genia. Genia was amazingly hospitable, and the house was often full of interesting visitors. One of the friends who came often to the house was Klaus Fuchs, then chief theorist at the British Atomic Energy Establishment at Harwell. Fuchs was a Jewish refugee from Germany who went with Peierls as a member of the British team to Los Alamos. First at Los Alamos and later at Birmingham, he was highly respected as a scientific colleague and also as a helpful baby-sitter. Rudolf and Genia both liked and trusted him. It came as a terrible shock to them when Fuchs was arrested and convicted in 1950 as a Soviet spy.

While I was still in Birmingham, I received a letter from Bethe with the news that Feynman had decided to leave Cornell and move to the California Institute of Technology. Bethe offered me Feynman's job, a full professorship at Cornell, beginning in September 1951, when my promise to the Commonwealth Foundation to remain in England would expire. I accepted the offer without much hesitation. It gave me what I wanted, a permanent place in the American scientific community. Already I felt more at home in Ithaca than in Birmingham. Bethe and Peierls were both first-rate theorists, but the experimenters at Cornell were doing far more exciting work than the experimenters at Birmingham.

In 1950 I received a dispensation from the Commonwealth Foundation allowing me to visit the United States for six months, provided that I returned to England for the rest of the promised two years. I arranged to spend the summer lecturing at the Ann Arbor Summer School, and to spend the fall at the Institute for Advanced Study. I made no secret of the fact that the purpose of this trip was to be reunited with Verena and to allow us to make decisions about our future.

Oppenheimer invited Tomonaga to come to the institute for the academic year 1949–50. He arrived after I had returned to England. I came back to Princeton in June 1950, one week before he returned to Japan. In that week I was able to get to know him as a friend. We talked at length about physics and also about his life in Japan. I always regretted having missed the chance to spend a longer time with him.

JUNE 24, 1950, PRINCETON

I have been here all this week and have had a pleasant time talking to people about the work they have been doing. I was lucky to meet Tomonaga, the third and most elusive of the Tomonaga-Schwinger-Feynman triangle. He is a charming man, like so many of the really good ones. He talked with me for three hours with much humour and common sense. On Thursday I was luckier still as I was travelling up to New York to say hallo to the Commonwealth Fund, and I found Tomonaga travelling on the same train, so we had another long conversation. That was the last I shall see of him as he now goes to San Francisco to take ship for Japan. He impressed me very much with his open-mindedness and quick grasp of ideas. He is more able than either Schwinger or Feynman to talk about ideas other than his own. And he has enough of his own too.

During this year he says he has not done a lot of work, he has used his time at Princeton as a holiday, and has greatly enjoyed the freedom from lecturing and running his department. Of course he is in Japan besieged with research students clamouring for help and advice, even more than I am in England. I have decided that I shall certainly spend a summer, sometime when I do not feel obliged to come and see you in Europe, visiting Tomonaga in Japan and keeping him in touch with the outside world. Incidentally, on his table when I went to see him was lying a copy of the New Testament. We had so much physics to talk about that I did not question him on that subject. But I have the impression that he is an exceptionally unselfish person.

July 11, 1950, Ann Arbor

I had a pleasant time the night before last. A string quartet was playing Haydn and Bach, and one of them was called away to a long-distance phone call, and they handed his fiddle to me and told me to take over. For fifteen minutes or so I played, and I was amazed to find I could still do it quite tolerably. I enjoyed it very much too. Perhaps I may start some serious practising and get back into regular playing.

August 5, 1950, Ann Arbor

I now have another letter from Sir John Cockcroft (director of the British Atomic Energy Establishment) offering me the Fuchs job (chief theorist) and telling me plainly that I am a bad boy for leaving

England and taking the Cornell post. This is rather unpleasant, naturally I have a bad conscience about the whole business. But I felt better after talking the situation over with George Uhlenbeck. Uhlenbeck described to me how he had come up against these two problems, (i) when he left Holland and came to America and (ii) recently when he was urged to return to military research (hydrogen bombs). He said on both occasions he was told by many people it was his duty to stay in Holland or to help with the bombs. He decided in both cases that there was not actually any question of duty at all. He would certainly help with the bombs if he were told it was his duty by President Truman or some responsible government authority, but not when he was told this by Edward Teller. He did not advise me directly but implied that I should look at Harwell in the same light. I should do the job if I felt I should enjoy being an important personage with a lot of files and correspondence and secretaries and committees and conferences, but not out of a sense of duty. Of course I am already committed to Cornell, and there is hardly a problem in deciding what to do. I wrote to Cockcroft to that effect.

August 16, 1950, Ann Arbor

The wedding was a simple affair. We had two witnesses, Helen Lennox for me and Professor Beno Eckmann, Verena's mathematics professor and an old friend of Zürich days, for her. Mrs. Uhlenbeck also came, so we were six all together including the judge who officiated. The judge said he was glad to have something to do because legal business was slack that day. For our honeymoon we went up to northern Michigan, a country of lakes and forests and long smooth straight roads. We drove many hundred miles, bathed and sat in the sun, and talked endlessly.

Today we have come through one of the difficult and delicate moments of our marriage, and we have come through it peacefully. Hans Georg Haefeli came for the afternoon to visit us. He brought with him a girlfriend, Inge von Richthofen, a relation of the (Red Baron) flier and a good friend of Verena. I was very glad to have finally met this man and made friends with him. He behaved admirably, indeed I have a great respect for him, and I think he is completely sincere in wishing us the happiness he could never achieve. Because of his tact and good management, the afternoon was not painful to us.

SEPTEMBER 5, 1950, BOSTON

Now we are here at the great International Congress of Mathematicians. There are about two thousand mathematicians here, and most of them seem to know either me or Verena. So we spend our days talking, being entertained and congratulated, and meeting old friends unexpectedly. It is wonderful how well we fit together into this mathematical society. For example, Daniel Pedoe, the man who introduced me to higher mathematics when I was at Winchester, is here, and he and Verena get on with each other like old friends. This is remarkable, because I had always found him rather a dull dog when I met him at Cambridge; but when Verena is around, he is scintillating, as he was when he and I used to walk the streets of Winchester. It is the same with Louis Mordell, the professor at Cambridge with whom I never could get on well in the old days. And I introduced Verena to André Weil, the great French mathematician whom I met in Chicago, and they were chatting together happily for a full hour.

OCTOBER 16, 1950, PRINCETON

I spent a week at Cornell staying with the Bethes. This was a good and successful visit. I gave one talk of a general kind to the whole physics department, one more specialized to the theoreticians, and about six hours of private talk with Bethe and two others, in which I explained fully my new idea of this summer. I felt very pleased with the whole affair. Perhaps after all they were not so foolish in choosing me for their professor. I came at a good time, with a new idea, and they gave me a royal welcome.

The new idea which came to me in Ann Arbor was to make a separation between high and low frequencies in the description of processes in quantum electrodynamics. The high frequencies could be effectively eliminated by Feynman's tricks, and the low freqencies would in the end contain an exact description of the process. The method would give us an exact description of nature, if and only if the series of high-frequency perturbations converged. I preached this gospel to enthusiastic listeners at Cornell, sincerely believing that the series would converge.

December 14, 1950, Princeton

Last night we had our farewell party, a small affair with eleven people. Tomorrow morning we start for New York to get the sailing permits and the reentry permits straightened out. And then Saturday morning walk onto the *Queen Mary.* Yesterday I managed unexpectedly to think out the solution of the outstanding problem which remained in my physics researches. This is a wonderful piece of luck. I will sit down and write another long paper as soon as we are settled in Birmingham. The new method is sophisticated, and people need a long time before they can assimilate it. People are respectful but also bewildered. This is true even of Oppy.

I arrived in January 1951 with Verena and Katrin for my second stay in Birmingham. Genya Peierls was enormously helpful in showing Verena how to cope with a English winter. Since Verena was pregnant, she moved to Zürich in April to prepare for the birth of a baby in July. I joined her in June and was invited by Pauli to work at the Federal Institute of Technology (ETH). Every afternoon while I was there, Pauli would invite me to walk with him around the city, talking about physics and about world affairs. He needed an audience, and I was delighted to be a listener.

In July Verena announced that the baby was on the way, and we had to walk fast to the Pflegerinnenschule to be there in time. We had a sharp rebuff at the door. The attending nurse told us that no child may be born in Switzerland without a boy's name and a girl's name chosen in advance. In Switzerland everything must be done in proper order. We quickly decided that the baby was either Oliver or Esther and were then admitted to the building. Esther made her appearance a few minutes later. She was a healthy baby with legal problems. According to Swiss law, she was my legitimate daughter and therefore British. According to British law, Verena was still legally married to Hans Haefeli, so Esther was illegitimate and therefore Swiss. Fortunately the United States was more generous than the other countries and accepted Esther as a stateless immigrant. She remained stateless for many years, with the consequence that she had to have a visa for every country that we passed through when we traveled abroad.

My parents came to Zürich for Esther's christening in the English church, with the consequence that there are no letters for this period. In September, Esther took the train with Verena to Genoa to catch a boat to America. There were big problems at the Italian border at Chiasso, where they were thrown off the train by the Italian officials because Esther did not have a visa. She was given an emergency visa just in time to catch the boat, driving overnight by taxi from Chiasso to Genoa. They had no problems with the immigration authorities at New York, where I was waiting with Verena's car to take them to Ithaca.

While I was in Birmingham, I had published in the Proceedings of the Royal Society of London *a paper explaining the new ideas which I believed would establish once and for all the mathematical consistency of quantum electrodynamics. When I published that paper, I believed much more, that the same ideas would provide a firm mathematical basis for the theory of nuclear particles. I believed that I was on the way to constructing an exact theory that would encompass the whole of particle physics. In Zürich I discussed my ideas with Pauli, and Pauli remained sceptical. Pauli said he could not prove me wrong, but he had a strong feeling that the crucial high-frequency series that I assumed to be convergent would actually diverge. One afternoon in Zürich, while I was walking with Pauli, I suddenly saw that his intuition was right. I found a simple physical argument showing why the series could not converge. I explained the argument to Pauli, and he said, "I told you so." I felt happy rather than sad to see my grand program so easily demolished. It was far better for me to have found the flaw in the program myself than to have other people find it later. Now I knew that the program was dead, I could write a short paper explaining why it failed and thanking Pauli for his help. My scientific reputation would be intact. I could recover from this failure and find new problems to work on at Cornell.*

NOVEMBER 2, 1951, ITHACA

Today it is snowing steadily. Yesterday it began, the first snow of the winter, and now it is nearly three inches deep. It is good that we are warm and dry in our little apartment. Esther is wonderfully healthy and well-behaved. She has not yet started to be noisy. She looks pretty, and her hair grows more blond as time goes on. I have written a short paper which I sent off to the *Physical Review* yesterday. It is about the ideas I worked out during the summer in Zürich. It is an unusual paper, with not a single equation in it, just some general discussion in the style of Niels Bohr. It is also original and perhaps not correct. I sent it off for what it is worth. It is rather like the famous paper of the Norwegian mathematician Niels Abel, proving that the general equation of the fifth degree cannot be solved. When Abel produced this about 1825, the mathematicians who had been trying for years to solve the equation were not pleased. Most of them did not understand the proof, and some of them said it was wrong. Now I have done the same thing to a lot of physicists, proving that what they have been trying to do is impossible, namely to get convergent series solutions to problems in electrodynamics.

DECEMBER 20, 1951, ITHACA

I think Verena already told you how I went off the road and wrecked her beautiful car in which we had had so many fine trips, both together and separately. I was driving to Syracuse to give a seminar talk there. On the icy roads I took good care and drove very slowly at corners and crossroads. But at the end as I came down into Syracuse, there was a piece of straight road, quite empty with no other car in sight, and so I put on speed to about thirty-five miles an hour. At that point there came a sudden strong gust of wind from the left, the road was a bit sloping down to the right, and the car started to skid and went off the road before I had time to do anything. With a tremendous bump, I hit a big wooden light pole, of the size and consistency of a telegraph pole. Luckily I was going by that time backwards, having spun right round, and so I was only thrown back into the soft seat and not forwards. I was completely unhurt. But the car was badly injured, and the people who looked at it agreed it would not be worth trying to mend it. I telephoned the University of Syracuse, and they picked me up and I gave my talk on schedule.

I was lucky not to be wearing a seatbelt when I crashed. I had time to stretch out on the seat with my head down before the impact. If I had been wearing a seatbelt, my head would have been above the seat back, and my neck might have been broken. It is true, as we are constantly told, that seatbelts save lives on the average. But if you happen to be skidding backward when you hit a light pole, you may be safer without a seatbelt.

10

CORNELL PROFESSOR

THE LETTERS in this chapter are from Ithaca unless otherwise indicated. In March 1952 came a big surprise, my election at the age of twenty-eight as a fellow of the Royal Society, the British equivalent of the American National Academy of Sciences. The Royal Society has a longer history and a ceremonial gravitas which the American National Academy lacks.

MARCH 24, 1952

Thank you for your telegram. That was a great joy to greet us at nine o'clock in the morning. I am glad they decided to elect me in spite of my having left the old country. Not so much for the honour and glory, though that is quite considerable over here, where the scientists hold the Royal Society in high esteem. I am chiefly happy about it because the Royal Society does an outstandingly good job in managing public affairs and political problems in which science is involved. The political organization of scientists over here is unsatisfactory, mainly because they lack a Royal Society. There is no organization which is not either politically negligible or dependent on the government. To have a body which is independent and still commands general respect is not easy. I hope I can either be useful to the

Royal Society as a representative over here or else keep in touch with them by periodic visits to England.

In 1952 my parents came to the end of their time in London. My father was director of the Royal College of Music from 1938 to 1952. When he retired, they returned to live the rest of their lives in Winchester, where they had lived from 1924 to 1938 and raised their two children. My father said it would be better for his successor at the Royal College if he removed himself from the scene. Winchester welcomed him back by giving him the Freedom of the City in a public ceremony. He continued to be active in public affairs. He was chairman of the Carnegie United Kingdom Trust, one of the largest British charities, from 1955 to 1960. He continued composing and conducting music occasionally until his death in 1964.

My mother was forty-three when I was born. She already had gray hair, more like a grandmother than a mother. She had a law degree and worked as a solicitor in her father's law office until she married. When I was a child in Winchester, she lived like a respectable proper English lady, careful not to let her friends know that she was running a birth control clinic in the town. When I was a teenager and we moved to London, she became my intellectual companion. We spent long hours together at the art galleries and museums and botanical gardens of London. We remained close until she died at ninety-four in 1975.

My sister Alice, three years older than me, also remained a close friend. She came several times for long visits to America. She was a medical social worker and worked in the Winchester hospital, a venerable institution with buildings designed personally by Florence Nightingale. Alice took care of our mother in her last years and continued to live in the family house after our mother died. Alice was a Catholic. She was well loved in Winchester, and a huge crowd of friends came to her memorial mass when she died in 2012.

My father's compositions were mostly choral works designed to be sung by amateur choral societies. His works were popular in Britain, where there is a strong tradition of amateur choral music. They were not well known in America, where popular music has different patterns, such as gospel, country, jazz, and rock. His best-known work was The Canterbury Pilgrims, *a dramatic setting of the char-*

acters described by Chaucer in his famous poem. "The Merchant" is one of the characters who is sung in joyful fortissimo by the whole chorus.

MAY 21, 1952

What a piece of luck that they should do *Pilgrims* for the first time, of all places in the United States, in Geneva, New York. Geneva is just forty-five miles from here, at the head of the next lake to ours. We were able to drive it comfortably in one and a half hours. If it had been any further away, we could hardly have made it, as it is the last week of classes and we are both tied up most of the days with students. When we arrived in Geneva, we first bought the local paper to find out where and when the performance was. I can say with certainty, it was a complete success. They had rotten luck with the weather, it rained heavily and continuously from ten a.m. till ten p.m. Still there was a respectable audience, about a thousand in a hall which would seat fifteen hundred. And they clapped like five thousand when the show came to an end.

I don't know what you would have thought of the performance musically. To my inexpert ear, it seemed that the chorus and the soprano soloist were excellent, the orchestra good, the tenor and bass soloists mediocre. The whole thing came across clearly all the time, it was never confused or dull, everything was lively. As we drove away from the hall after the show, we passed five or six girls of the choir in their long evening dresses, walking home and singing "A Merchant Was There with a Forked Beard" at the top of their voices. That was a beautiful finale. We decided not to introduce ourselves to the performers, partly from shyness and partly because we wanted to get home by midnight. But I wrote Mr. Lafford (the conductor) a note of thanks and congratulations this morning, saying who I am. About the city of Geneva, I know nothing, except that it is about the same size as Ithaca and has a flourishing beer-brewing industry. And I regret to say that until yesterday I had never heard of Hobart and William Smith Colleges. Evidently Mr. Lafford is an enterprising fellow.

AUGUST 27, 1952

Verena was invited by one of her classmates from Zürich University, a girl called Edith Müller, to go and stay for a weekend with her in Ann Arbor. Edith has made a good career for herself as an astronomer, and she is now working at the Ann Arbor observatory for two years as a research assistant. She avoided getting married, and she lives in a big comfortable apartment by herself, and she loves her work. So Verena was glad to accept the invitation. Ann Arbor is a beautiful place, and I was happy to walk around and revive old memories. The only trouble with this little holiday was that we had to drive 550 miles each way. We had intended to go through Canada, which is the quickest and quietest way. However, we ran into trouble at the border. They would not admit Verena and Katrin into Canada because their papers are not in order. When this was discovered, we were in Canada, since the office is on the Canadian side of the bridge. So the officials had to give us an official order of deportation to get us back into the United States. That is the first time we have ever been deported. It makes us feel very distinguished.

SUNDAY, SEPTEMBER 7, 1952, BROOKHAVEN NATIONAL LABORATORY, LONG ISLAND

I have been living here at Brookhaven for six days. It is a great time to be here, as the lab has recently made a new discovery of major importance. The discovery was made by two people, one an old friend of my own generation called Ernst Courant whom I used to know in Cornell in 1948, the other a professor from MIT with the unlikely name of Stanley Livingstone. Courant is on the regular staff here, Livingstone was here for the summer. The discovery does not sound spectacular. It is a new way of building a magnet so that a proton or an electron can be guided through a narrow tube without hitting the walls. But this rather simple idea will completely change the economics of high-energy particle physics. Also it will have many other uses as time goes on.

Here they have a big proton accelerator called the cosmotron, which Livingstone and Courant designed four years ago. It is just now starting to work. They got protons from it for the first time in June of this year. The protons had an energy of two billion volts. The cosmotron cost $7 million, and this is about the limit of what it is thought reasonable to spend on such machines. It is also well beyond the limit of what any university laboratory either here or abroad can do. So the effect of economic facts has been to concentrate important experiments more and more into government labs and away from universities, also into the United States and away from the rest of the world, two tendencies which everybody deplores.

This new idea of Livingstone and Courant makes it possible to build much more powerful machines for the same cost as the cosmotron. Here they have the preliminary designs for a new machine which will again cost $7 million, to be ready in about four years and giving protons with thirty billion volts. These protons will go round and round a circular steel pipe which will be six hundred feet across the circle but only one inch in cross-section. I have looked at the mathematics, and it seems it will work. The machine will certainly be built, and it will certainly be needed. All it lacks yet is a good-sounding name like the cosmotron. It has been a great pleasure to me to be in at the beginning of this new idea, to see how it has developed. It was only because these people have spent four years building a big machine that they suddenly had the idea how to do it better. Such ideas do not come by abstract thinking about the problem.

The new idea was called strong focusing, and it was immediately adopted by builders of particle accelerators all over the world. The new machine that was built at Brookhaven was called the alternating gradient synchrotron. It was for many years the leading instrument for high-energy physics experiments.

SEPTEMBER 24, 1952

Today is the first day of term, and the university feels gay with students swarming everywhere. They look absurdly young and cheerful. I have my first class to teach tomorrow and Verena has hers today. She is teaching calculus to the freshmen (first-year undergraduates) which will not be any intellectual strain for her. She will have a fair amount of work to do correcting the homework, but it will be easier than the differential equations she had last year. She feels happy to be doing any kind of a job, and she also keeps up some contacts with the maths department. My course is Introduction to Quantum Mechanics, also on a more elementary level than last year's work. This should be an interesting course to teach, the first introduction to quantum theory being the critical step in the education of every physicist, the first time the students meet new and difficult ideas, the first time they are forced to think hard. Usually they go through a stage of complete confusion and depression as a result of meeting the quantum theory. Probably this is unavoidable. It will be fun to see how it goes.

OCTOBER 25, 1952

Now for our main news. Two weeks ago I made up my mind that this life of being a professor with a lot of students to look after, combined with a growing family, is taking up too much of my energies, so that it is not possible for me to do much in the way of serious thinking. I decided that I ought to do serious thinking during the next few years, when I have a chance of making important contributions to science. So I decided to write to Oppenheimer and ask him if he would give me a permanent job at the institute at Princeton. Several times in the past he has said casually he would like to have me with him. So I wrote a carefully worded letter asking what he could offer me, laying all my cards on the table and explaining my difficulties.

Two days later there came a long-distance call from Princeton, Oppy on the line. He said, "Certainly I will get you a permanent position at Princeton if you want one. Only it will take a little time before I can get a formal offer approved by my committee and sent out to you." So it seems definite that we shall move to Princeton, either in fall 1953 or fall 1954. The exact terms of the offer I do not yet know. But I presume they will be satisfactory.

We are both happy about this change. First, it means I will have some mental peace and the best conditions for working. Second, the Princeton climate is much more agreeable. We already had snow which stayed on the ground two days, it is freezing hard tonight, and we now have a solid six months of winter to look forward to. In Princeton the winters are not so exhausting. Third, the friends Verena has in Princeton are closer and more desirable to her than any she has here. This means a lot in the long run. I also find the same thing true of myself. Hans Bethe is the one man here with whom I really feel at home, and he is here so little of the time and he is so busy that I do not see much of him.

By an unhappy coincidence of which I was unaware, I arrived as a professor at Cornell in September 1951, just at the time when Edward Teller and Stanislav Ulam discovered the trick that makes hydrogen bombs feasible. The secret program to develop hydrogen bombs at Los Alamos changed suddenly from a leisurely stroll to a furious sprint. Hans Bethe was called in to help with the design of the Ivy Mike thermonuclear device, which was built in a hurry, tested on November 1, 1952, and exploded with a yield of ten megatons. So it happened that Bethe was away from Cornell for much of the first year after my arrival, and I did not know where he was. He was missing just when I needed him most.

Since I wrote to Oppy, I find there is one more complication. Unknown to me, Bethe himself had recently also been offered a permanent position at Princeton, and he is seriously considering accepting it, for much the same reasons as I. This actually does not make much difference to me. If he goes, it will only be better for me in Princeton and worse here. All the same, I somehow doubt if he will go, he is so much a part of the landscape here, and he feels strongly rooted too. I have no doubt that if I had to stick it out here, I would manage to make life easier for myself and should end up being happy and able to do good work. But since I have this opportunity of escaping, I may as well make the most of it.

Two days after the phone call from Oppy, I had a conversation with Bethe, and out of it I got a new idea for a major piece of research in physics. This has absorbed me completely during the last ten days. It goes ahead well and already has led to good results in understanding some recent experiments in the properties of certain particles called mesons. I have been able to find jobs for some of my students to do in connection with this idea, and this makes them happy too. I have hopes this will lead in a few weeks or months to a big step forward in the clarification of the whole meson theory. These last ten days have been great fun, but they make it even more clear how necessary it is to go to Princeton. I simply cannot go on at this pace. All my routine jobs are left undone and are piling up ahead of me. Still it is good to have such work to do, and in time I shall get it done in spite of exhaustion. Bethe is wonderful, if only I could see more of him and less of the students. We definitely go to California for the summer. That is June to September. I am teaching a summer school in Berkeley. This will be a fine change of air for Verena. We shall go all together and rent a house for the summer.

November 1, 1952
[THE DAY OF THE IVY MIKE TEST]

The decision is now definitely made that we leave here and go to Princeton. Only it is a question whether they can find a successor in time for us to leave next year. The successor is likely to be Ed Salpeter, who is already here as a research associate and does Hans Bethe's lectures for him on the frequent occasions when Hans is away. Ed is a very good man, and I think they could not do better. I am happy that he is here. It makes me feel less compunction about leaving Cornell that they have such a good man to put in my place. This is still completely unofficial. Ed Salpeter is an Australian originally from Vienna who took his degree with Peierls in Birmingham and came here three years ago. He is not a deep thinker, but he is a productive worker and a good teacher and above all physically and mentally tough, so he should be suitable for this job. He and Hans Bethe are good friends too. He has married a Lithuanian-Canadian girl who was a student here and is now taking a degree in animal psychology. The question whether Hans Bethe will leave is not to be decided for several months. I believe he will not go.

I have been busy this week. It goes well with my new ideas and the calculations connected with them. Now I have about six people here working on this job, including Salpeter. I am rapidly reaching the point where I only talk and the others do the work. I hope it will not stay like that—I need to get back to real work myself if we are to keep on making progress. The most remarkable thing is the way people outside are already beginning to get excited about this work. On Wednesday I had a phone call from Princeton from a young Chinese physicist called Lee. He announced he would arrive on Thursday by air in order to learn what we were doing. He duly arrived, and I spent a day telling him all about it. He is a charming young man and understood everything readily.

Tsung Dao Lee was a brilliant young theorist who had been a student of Fermi in Chicago with his countryman Chen Ning Yang. Lee and Yang together won the Nobel Prize in Physics in 1957 for their revolutionary theory of weak interactions. Lee spent most of his life as a professor at Columbia University in New York. In later years he was given a royal welcome in China and organized a program for bringing Chinese students to study physics in America. After one of his visits to China, he told me with pride that he had survived thirty-one consecutive banquets.

Esther is the healthiest baby I ever saw, now that the cold weather is back and her cheeks are red again like ripe apples. But she still doesn't talk. Katrin is growing up fast and beginning to get more helpful around the house.

My activities in the department are growing by leaps and bounds. I now am directing an empire of eight people who are working hard on the meson calculations which I started six weeks ago. It is amazing how things are humming. Everyone is happy, and they are getting interesting results. It is easy to run such a group once you have a suitable job for them to do. I am happy about it all. When I leave here, they will say "Look how he built the department up in two years" instead of "He didn't like it so he quit after two years." This makes a great difference.

The big program of calculations of meson scattering that I had organized at Cornell was intended to explain the big program of experimental measurements of meson scattering organized by Enrico Fermi at Chicago. I took our theoretical results to Chicago and showed them to Fermi. They agreed quite well with his experiments. But Fermi was not impressed. He said, "How many input numbers did you use to fit the experiments?" I said, "Four." He said, "As my friend John von Neumann likes to say, with four input numbers I can fit an elephant." For Fermi, the numerical agreement meant nothing. He said politely that our calculations were worthless because they were not based on a well-defined theory. His intuition told him that our description of nuclear processes by a set of equations borrowed from quantum electrodynamics had missed something essential. Fermi died two years later, long before the missing ingredient in our description was discovered. The missing ingredient was quarks. I returned from Chicago to Ithaca to tell the students the sad news that our whole program of meson scattering calculations was a grand illusion.

DECEMBER 17, 1952

In the evening there was a big party of the whole physics department, students and staff, about two hundred people. It was not a birthday party for me but the regular annual Christmas party. However, it turned out to be something special. They gave an entertainment which consisted of a representation of a qualifying examination, an oral exam which each Ph.D. student has to take individually before three professors. The professors are always one theoretician, one experimenter, and one mathematician. Each professor has to do these exams rather frequently since there are a lot of students. I do it about once a month. The play was in two parts, first the exam as seen by the professors (with an offensively stupid student and very patient professors) and then as seen by the student (with a clever student falling into traps laid for him by the malicious professors). It was all very funny. But the funniest thing for me was that the theoretical professor spoke with a pure English accent and said "Bad luck" when the student made mistakes, and started to eat a sandwich halfway through the exam. He was unmistakably intended to be Professor Dyson. This was for me completely unexpected, and it is good to find that I have already become enough of a public institution to have such jokes made about me.

DECEMBER 30, 1952

I came back from the Rochester conference with two Japanese physicists, Yoichiro Nambu and Toichiro Kinoshita, who stayed with us overnight, and the next day went out with me and Katrin and helped cut down and carry home the Christmas tree. They were interested in the quaint customs of the Westerners. They have been over here only four months and are at the Princeton institute. I had long talks with them about physics and learned a lot.

I was lucky to get to know these two Japanese geniuses when they were still young and not yet famous. Both of them stayed in America and became central figures of the theoretical physics community, Nambu in Chicago and Kinoshita at Cornell. Nambu won a belated Nobel Prize in 2008 for his theory of broken symmetry. Kinoshita spent his life pushing the calculations of quantum electrodynamics to higher and higher accuracy, barely keeping pace with the increasing accuracy of experimental measurements. In 1952 the theory and experiment agreed to an accuracy of two places of decimals. We all then expected that discrepancies would arise when theory and experiment would go to higher accuracy. In 2013 Kinoshita came to my ninetieth birthday celebration in Singapore to report on the latest comparison, which showed theoretical and experimental results still in agreement to an accuracy of eleven places of decimals. It still seems miraculous to me that our makeshift theory, which we expected to be quickly superseded, is still alive. After sixty years of rigorous testing, Nature still dances to our tune.

In the letters, there is no mention of any discussions with our Japanese visitors about the bombing of Hiroshima and Nagasaki. In those days there was a general agreement on both sides that the bombing had saved lives by bringing the war to a quick end. We all had vivid memories of the war and felt that we were lucky to have survived it. Serious questioning of the ethics of the bombing arose later, after the joy of surviving the war had faded.

FEBRUARY 3, 1953

I had a letter from [Maurice] Bowra, vice chancellor of Oxford, offering me the chair from which Sydney Chapman is retiring. You may remember Chapman was my boss at Imperial College in 1946. It is a fine offer, and in many ways both Verena and I would be glad to go. I think if I had not this new job in Princeton, I would have had a hard time making up my mind about it. Oxford would offer just the things we lack here, a more international society, contacts with Europe, longer vacations, and more leisure. However, now that Princeton offers us all the same things and double the salary, we decided without much difficulty to say no to Oxford. Also we both feel appalled at the thought of once more packing up all our stuff to cross the Atlantic. I think I may one day come back and take such a position at Oxford, but it will not be before the children are grown up and able to take care of themselves. Say in twenty years from now. If they still want me.

On March 26, 1953, our son George was born at the Ithaca hospital. Another healthy baby, this time without legal problems. It came as a great relief to have him born in a country that gave him citizenship as a birthright, even if he was an illegitimate child of two aliens.

APRIL 1, 1953

Since Katrin has no school this week, she is allowed to stay up later. The three of us sat over our drinks and discussed the world and its problems. Katrin is coming now for the first time into contact with these questions. It is beautiful to see how she grasps and reacts to them. This evening she started the conversation with a little speech. I do not know where this came from. I did not talk to her about such questions because I was full of the new baby and our family problems. She started off saying, "But it's true, isn't it, Freeman, that the world is in bad shape? The whole world, I mean, not just America and Switzerland and stuff like that. There are so many people everywhere who are bullies, and there are so many wars all the time, and there are all kinds of things that hurt people. I don't see why there should be so many things that hurt people, do you, Freeman? I think I should like not to kill anything, never to kill anything, not even flies and stuff like that, and chickens and rabbits. And then maybe the bees in the garden wouldn't want to sting us anymore. And the people in different countries could all be friends with us, think how fine it would be, we could all belong to one tremendous big family, and think how many babies we should have to look after, what a big family it would be. And maybe even the lions would get friends with us and we wouldn't be afraid of them anymore." All this came out suddenly, out of the blue, and I made an effort to remember it and write it down just as she said it. I love Katrin more and more as time goes on. She has so much tenderness and imagination, and her life is not easy being the eldest in the family.

MAY 23, 1953

I am giving my final lectures, ending one week from now, and setting exams, settling the affairs of my students. I enjoy all of it, because this is the last time I do these things. I had yesterday two surprise invitations. One was to serve as assistant editor of the *Physical Review*, of which there are about twelve; it is a tedious job, but I accepted as a public service. The other was to give two weeks of lectures about physics next year at Harvard. Both these are things I am well fitted for. It is a great advantage of my going to Princeton that I can take on such incidental jobs without difficulty. If I were at Cornell, I could hardly go to Boston for two weeks in term time, and the *Physical Review* job would be only a burden. But in Princeton I can do such things and have enough time over for myself as well. I feel the move to Princeton will not only be good for my natural laziness but will also be the right thing for increasing my usefulness to society. Anyway let us hope so.

11

MYCENEAN TABLETS
AND SPIN WAVES

*In Berkeley, California, there was a lively group of solid-state physi-
cists with experimenters and theorists working together. Arthur Kip was the leading
experimenter, Charles Kittel the leading theorist. I wanted to move out of particle
physics and try my hand at something new, so I arranged to spend the summer of
1953 working with Kip and Kittel. I taught quantum mechanics to a class of summer school students and meanwhile taught myself solid-state physics by working on
problems suggested by Kittel. The whole family came to Berkeley, and we rented a
beautiful old house on the steep hillside above the campus.*

AUGUST 14, 1953

I am getting a lot of good physics done since my lectures stopped.
I sent off one paper to the *Physical Review*, and I am now deep in
another one. This latest piece of work is in connection with the solid-
state project which is paying part of my salary while I am here. Fortunately they had some problems which I was able to solve, so I am
earning my keep. It is good that they get value for their money, so I
will find it easy to get invited here again.

*The main problem that I solved was to explain Arthur Kip's experiment on
electron spin resonance in metals. Kip put a piece of metal into a cavity filled with*

microwaves at a fixed frequency with a variable magnetic field and measured the absorption of microwaves by the metal. He found an unexpectedly sharp resonance, which I was able to explain as a consequence of the dynamics of electron spins in the metal. The microwaves only penetrate into a thin skin on the surface of the metal, and the electrons in the metal see the microwaves only for brief snatches of time, but the electrons remember the phase of the microwaves from one snatch to the next. The accurate phase memory of the electrons makes the resonance sharp.

In September 1953 I started my career as professor at the Institute for Advanced Study. We lived for three years in a house rented from the institute, then in 1956 bought the house that became our permanent home. The main change at the institute since my previous visit in 1950 was the Computer Project, bringing together engineers and scientists to build the machine and use it. In 1953 the project was in vigorous operation. My closest friends were the two people who were leading the project, Julian Bigelow, the chief engineer, and Jule Charney, the chief meteorologist. John von Neumann was nominally in charge, but his work designing the machine was finished, and he was spending most of his time in Washington as a member of the U.S. Atomic Energy Commission. I looked forward to a future with the institute leading the world in two new sciences, computer science led by Bigelow and climate science led by Charney. My hopes were quickly crushed. Within two years, von Neumann was dying of cancer, and the institute had decided to close the project down. I was the only member of the institute faculty who considered the closing to be a disastrous mistake. Oppenheimer was not interested in keeping the project alive, and I could do nothing to save it.

At that time Einstein was still alive, but he was never interested in the Computer Project. Einstein was also uninterested in particle physics. Since the young physicists were mostly working on particle physics, Einstein took no part in our activities. We saw him each morning walk from his home to the institute, and each afternoon walk back, but we never spoke to him.

OCTOBER 1953

We have meetings which I attend as a member of the faculty, about institute financial policy, new appointments, and so on. I find these absorbing, and I enjoy Oppenheimer's ability to understand

and explain everything. Next Wednesday Oppy will go for a month to London and will give the Reith lectures for the BBC. I hope you will listen and see what you think of him. He has put a lot of work into preparing these lectures. Maybe you might even run into him at some function or other.

NOVEMBER 20, 1953

A political struggle agitated Princeton. A referendum vote was taken to decide whether or not Princeton should be "consolidated." This was an instructive affair and showed the vigor of local government in America. Princeton consists of two separate entities, Princeton Borough and Princeton Township. The borough consists of the old town and has a population of nine thousand. The township consists of the suburbs and has a population of six thousand. They have separate governments, separate schools, and separate taxes. So some busybody reformers decided we ought to consolidate and distributed a lot of propaganda to this effect. In reply, the shop windows blossomed with posters like this: WE HAVE SAVED PRINCETON FOR 150 YEARS. SAVE PRINCETON NOW. VOTE NO TO CONSOLIDATION. Finally, when the great day came, we defeated the radicals. Both borough and township voted against consolidation by three-to-two majorities. We are happy about this outcome. The schools in the borough where we live are good and not overcrowded, and the population is stationary. The township has a rapidly growing population, and the schools are more crowded every year. So we are happy not to have to share.

A week ago I took a day off and Katrin had a school holiday, and we went to New York for the day. We went first to the tip of Manhattan and took a ferry to Bedloe's Island, which is a fifteen-minute ride with magnificent views of the shipping and the city. It was a cool day, and we ate hot dogs to keep us warm. On Bedloe's Island stands the Statue of Liberty which I had never seen at close quarters. We walked up inside her some 330 stairs. This is great fun as she consists of a copper skin with a steel framework inside and the staircase climbing up the steel framework. As you go up, you can see her anatomy from the inside. Finally you go up past her nose and eyes and stand on a platform level with her eyebrows. There is a row of windows along her forehead just under her hair, and you look out of these across the harbour, or upwards at her right hand with the torch, or downwards at her nose and her feet. It is a delightful place. There is also a gallery on the outside at the level of her feet, which is halfway up the whole monument, where you can walk round in the open air. After this we took the ferry back to Manhattan. On Sunday when we went for a walk, Katrin said to me, "Why don't we have a practice earthquake sometime at home?" I said, "What do you mean?" She answered, "Well, we are always having fire drills in school."

JANUARY 12, 1954,
PHYSICS DEPARTMENT, HARVARD UNIVERSITY

Everything goes well here. Last night I went to the Society of Fellows, a group modelled on the Cambridge College system. There are about fifty fellows attached to the society, and they come from all fields. We had a dinner such as Trinity College would have provided twenty years ago, lots of silver plate and a little trolley for pushing the port wine up and down the table. I had the good fortune to sit by Professor Hisaw who is a famous character of the university. He is a biologist and said to be the finest lecturer in the university. He comes from the hills of southern Missouri, the legendary home of American hillbillies and amiable criminals. He is evidently a born showman. We talked mainly about psychical research, as I had been reading some article on the subject by Aldous Huxley appealing for a more open-minded appraisal of the evidence. Hisaw said yes he was very open-minded about psychical research, so much so that he had served for five years on the "Spook Committee" of Harvard University. This committee was set up to administer a bequest of $15,000 a year which some good lady had left to the Harvard Corporation for the support of psychical research.

The committee consisted of professors from various fields, and one lawyer to see that the terms of the bequest were complied with. The committee invited anybody to submit proposals to them for examination, and they had proposals from all over the country which they investigated with great care. Every kind of crackpot came crowding for a share in the money. So the committee spent their days giving interviews, watching dice-throwing experiments, and arguing interminably. After five years they still had not found any applicant who deserved their support. But there was a professor at Columbia whose proposals seemed to be a bit better than the others. They finally decided (1) that the Columbia professor get the use of the money for his experiments for five years, (2) that he report his accounts and his experimental results to the lawyer who should be solely responsible for the execution of the bequest, (3) that the committee be discharged.

The book by Aldous Huxley that I had been reading was The Doors of Perception *(1954). Huxley did not succeed in making psychical research respectable. His writings had the opposite effect, blurring the distinction between scientific investigation of psychical phenomena and recreational use of psychoactive drugs.*

I met an interesting character at breakfast at the Faculty Club. He is president of a big aircraft manufacturing company, a man of about sixty. Oppenheimer happened to drop in for breakfast, and so I was introduced to him. This man had been travelling round the world making business deals, and had a very poor opinion of the industrial management wherever he went. He said, "People are surprised that the Italian workman votes Communist. By God, if I were an Italian workman, I should vote Communist. That is the only party which says cut the boss's throat. And that's obviously the only thing that will do any good." He also talked to Nehru. Nehru said it would be a fine thing for India to have nuclear reactors to power the new factories, because India has such poor roads and railways and they cannot transport enough coal from the coalfields. So he would like America to supply him with a reactor or two. This American said, "I just couldn't make the man understand that it's no good having factories if you don't have the roads to take away the stuff the factories make."

Tonight I shall go with Hans Haefeli to watch the big ice hockey game between his university (Boston College) and mine (Harvard). That will be fun too.

The match between Boston College and Harvard was a hard-fought battle. Harvard had better skaters, but Boston College had better teamwork. Boston College won. We were sitting on the Boston College side of the rink, and immediately behind us was a long row of Catholic priests shouting in unison. Not only the players but also the spectators had better teamwork on the Boston College side. I still treasure the memory of that game as an epic encounter of the two cultures coexisting in Boston, the old aristocratic Protestant culture of Harvard and the new team-oriented Catholic culture of Boston College.

MARCH 13, 1954

Since we came back, the main interest has been a talk at the institute on "The Deciphering of the Mycenean Script" by Professor Alan Wace, an English archaeologist. This is a fascinating story, maybe you also have read about it in the papers. In the Cretan civilisation which was destroyed about 1400 B.C., there were three kinds of writing at different periods. Over the last fifty years they have found about two thousand clay tablets, most of which belong to the latest period and are written in B script. I remember being taught in school that these writings would probably never be deciphered, because all trace of the language they were written in had disappeared, and the Greeks who came in from the north about 1400 B.C. had destroyed the native cities and introduced their own language during the following centuries of barbarism. Now during the last five years the whole picture changed because it was found that the Mycenean civilisation on the Greek mainland was using the same B script in many different places, during a period 1500–1200 B.C. which overlapped with the B script in Crete, and continued until the mainland cities were themselves destroyed by later invasions. In particular about one thousand of these clay tablets were found in the so-called Palace of Nestor at Pylos. So it became clear that the later Cretans were speaking the same language that the mainlanders were using right up to the Homeric period, and this could hardly be anything else but Greek. The archaeologists now consider that the original Greek tribes moved down the mainland about 2000 B.C. and reached Crete about 1600. The older Cretans had their own language and their A script at that time. The Greeks then learned to write their own language, inventing B script for the purpose, and took this art back with them to the mainland.

With these new ideas two gentlemen in London [*Michael Ventris and John Chadwick*] sat down to decipher the B script, assuming it to be an old form of Greek, and the experts are convinced they have succeeded. The script has eighty-eight signs, each representing a syllable. They worked out by trial and error phonetic equivalents for sixty of the signs, the other twenty-eight being so rare that they are not yet sure what they are. In this way they found the majority of the tablets make clear sense; they are mostly shopkeepers' accounts and inventories. The decisive check on all this came when some completely fresh tablets arrived from the excavations in Pylos, which had not been used for the analysis. One such tablet was a list of pots, giving on each line a description in words, a number, and then a picture of the pot. The first line was

TI-RI-PO 1 [sketch of pot with three legs]
TI-RI-PO-DE 2 [sketch of pot with three legs]
KE-TO-RO-PO-DE 2 [sketch of pot with four legs]

and so on. All this is enormously exciting to the classical scholars around here, now suddenly to have Greek texts from two hundred to five hundred years before Homer. They say they expect to find lots more of these tablets as soon as they dig for them. The first to be deciphered are the shopping lists, etc., which are easiest and least interesting. They already got one from Pylos which seems to be instructions for making a sacrifice to Poseidon, and more interesting material will become accessible as they learn more about the language. They say also that if they can learn enough about the B script, then they may be able to go back to the A script and discover what the original Cretan language was. For this they would need some kind of bilingual tablets.

For the general public who are not expert in classical languages, the most striking aspect of the Pylos discovery is the fact that the word tripod *is preserved all the way from Linear B to modern English. For the experts in classical languages, the most striking fact is that the Linear B word* ketero, *meaning "four," is closer to the Latin* quattuor *than to the classical Greek* tetra. *For this word, the Latin language has deeper roots than the Greek.*

MARCH 30, 1954

I am having trouble with the British authorities here to get Esther accepted as a British subject. I started with the paperwork last December, but it is not half done. People talk so much about the stupidity of the American immigration officials, but I never met anything so bad as these British. Every time I write to them or telephone them, they want more documents and more irrelevant information. After they finally decide that Esther is a legal child, they have to send the documents to Zürich to get her birth officially reregistered there, and then after all that comes back again, we can start getting her a passport. God knows whether all this will be done before May 26. Verena is also having trouble with her Swiss consul getting herself a passport. This may hold us up too. The only one who has no trouble so far is George, who is a bona fide American and gets an American passport.

We decided now we shall all become Americans as soon as we are eligible (1956) and have an end to these complications.

APRIL 21, 1954

After four months of delay, we finally today got a note from the British consul that he will not recognize Esther and George as British subjects and will not give them passports. This makes me so raving mad, I do not know what I may not do.

Oppenheimer is still in Washington having his life history examined by the Personnel Security Panel. We expect their verdict in about a week. We have been expecting this case to come up for a long time, and so far it must be said the government has handled it with unusual decency. It all depends now on the verdict. If the verdict is unfavourable, it means a major loss of contact between the government and the people who are supposed to advise the government on scientific questions. This will be bad in the long run. But it will not change anything here at the institute. I just came back from Washington where I attended the three-day physics conference. I took down a huge parcel of clothes for Oppenheimer and his family, which I delivered to the office of his Washington lawyer. His actual whereabouts is kept secret, so he and the family can get quiet evenings without being bothered by newspapermen. Bethe saw him in Washington and said he looks thinner but otherwise in good shape.

MAY 9, 1954

I wrote a strong letter to the British consul in New York, instead of my usual polite requests, and this produced some effect. Though it does not change the facts of the situation, the consul decided to give Esther a passport valid for a year, not renewable unless her status is established in the meantime. From a practical point of view, this is all we need for this summer. The consul, after I had been rude, did exactly what he said was "absolutely impossible" when I was polite.

The facts are these. In 1949 Hans Haefeli divorced Verena in his hometown of Balsthal, canton Solothurn, because according to Swiss law he remains for his whole life legally attached to his place of origin. He was then already resident in Massachusetts. According to British law, this divorce is valid if it is recognized as valid by the Commonwealth of Massachusetts, but not otherwise. The British consul was told by his lawyers that Massachusetts probably would not recognize the divorce, so he does not recognise it either. Verena and I are therefore not married, our children are illegitimate, and hence not British subjects. The joke is that we heard from Hans Georg when he was here that his lawsuit in Rome has finally succeeded, so the pope has agreed to the marriage being annulled. According to the pope, Verena was never married to Hans. The Swiss government however recognizes both marriages and the divorce, so according to them Esther and George are British.

MAY 23, 1954

Professor Wace, who stimulated my interest in the Mycenean inscriptions, is travelling to Europe the same day as we but on a different ship. He is going to Mycenae to dig, as he did last year, and intends to go on for many years more. I never ventured to ask him his age, but when we went together to Philadelphia, the income tax clerk asked it of him. He said in a sepulchral whisper, "Don't tell anybody. I am seventy-four."

I spent most of the summer of 1954 at Cécile DeWitt-Morette's summer school at Les Houches, with Georges Charpak as my star student. After that I went to the International Congress of Mathematicians at Amsterdam. Verena left the children in a Kinderheim in Switzerland and spent the summer visiting her sister who was living in Damascus, Syria. In those days Syria was a peaceful country friendly to expatriate Europeans. In September 1954 the family was reunited in Princeton.

OCTOBER 18, 1954

You probably have seen in the papers that Robert Oppenheimer got officially reelected to his job as director by unanimous vote of the trustees. So our anxieties about this are over. Everything here is now calm and back to normal. I have been spending the evenings reading the full transcript of the testimony at the hearings. It is a book of 992 pages of small print. I find it so enthralling that I don't mind the small print. It lays everybody's soul bare.

After the government decided to deny Oppenheimer's clearance, there was some speculation in the newspapers that the institute trustees might also decide to fire him from his job at Princeton. If he was not to be trusted with government secrets, the trustees might decide that he was not to be trusted with institute secrets. The institute faculty, including me, signed a public statement that we had confidence in Oppenheimer and wished him to continue as director. I was prepared, in case Oppenheimer was fired, to resign from my position at the institute and return with my family to England. It came as a big relief when the trustees made their statement reelecting him as director. We did not need to drag our children back across the ocean.

DECEMBER 15, 1954

Esther now is growing up fast. It seems always the rule that the children are most difficult when they are going through some mental development. She is now happier again and with a much wider range of ideas. Esther is fond of the *Alice in Wonderland* style of conversation. This is a conversation which is brought to an abrupt end by a remark which is at the same time entirely logical and entirely absurd. A good example is this. Esther, "I need a new head." Freeman, "What do you need it for?" Esther, "For my neck."

JANUARY 17, 1955

I was invited by the Soviet Academy of Sciences to a meeting in Moscow, to discuss the branch of physics in which I am expert.

The dates are March 31–April 7. It seems to be a genuine and serious affair. I decided I would make a fight and find out if it is possible to go. I think the chances are small, but one ought at least to try. So I have been busy writing letters to various officials, British and American. As our term ends April 10, there would be no hurry for me to get back to Princeton, but I might get a chance to travel around in Russia or elsewhere in Europe after the conference. In the newspapers today I see that Eisenhower approved a report of the National Science Foundation to Congress in which they said, "We must find out more about what the Soviet scientists are doing." This perhaps improves my chances.

FEBRUARY 4, 1955

I came home last night from a strenuous week of conferences. My Indian friend Abdus Salam (who is now fixed in Cambridge) flew over for the meetings. Dick Feynman was there from Pasadena. As far as we can discover, Salam and Feynman and I are the only three who have invitations to Moscow. We are all three seriously trying to go. It seems I am the furthest along so far. I had a prompt reply from the British consul in New York saying he has enquired from London, and HM Government will give me all the usual diplomatic protection and has no objection to my making the trip. I also wrote to the American immigration authority demanding a written statement that they approve this journey and will let me back into the country afterwards. If I do not get their approval in writing, I shall not go. The situations of Feynman and Salam are different. Feynman is a U.S. citizen by birth and so could have no difficulty reentering the country. But he has worked at Los Alamos, and so they may not feel inclined to give him a U.S. passport valid for the USSR. Salam has a more complicated problem because he is not resident here but wants to visit here from time to time. He wants the American visa authority to approve his going to Moscow so that they will not hold it against him at some hypothetical future time when he may apply for a visa. I think that of the three of us, I am in the most favorable position, and it may well be that I am the only one who succeeds in going.

All through the years of war and political hostility, some personal contact between Soviet and Western scientists was maintained. Some senior scientists from the USSR were permitted to travel to the West on official business, and a smaller number of Western scientists were admitted to the USSR. We knew that there were many excellent physicists in Russia who were not allowed to travel, for example Pyotr Kapitsa, Igor Tamm, Lev Landau, Andrey Kolmogorov, Vladimir Veksler, and Yakov Zeldovich. We could read many of their theoretical papers in the open literature, but they published little about experiments. The invitation in 1955 offered me a chance to make friends with the Russian theorists and perhaps to find out what the experimenters were doing.

March 1, 1955

I discovered some interesting things in the latest issue of the Russian physics journal. There is a string of seven papers describing experiments they have done with a big proton accelerator of 660 million volts. They have had the machine operating for three years, but until now there has never been a word published about any experimental work in nuclear physics. Of course, it was idiotic to keep such a thing secret. (American machines of this kind have never been secret.) It shows some glimmerings of common sense that they now make it public. My trip to Moscow may be more profitable if they are free to talk about their experimental work. What they have done with the machine is competent and careful work, but nothing particularly new or exciting.

March 18, 1955

I am sorry to say that I will not come and see you in April, as the Moscow scheme failed. The crisis came on Monday, when I went to

Washington to the central office of the Immigration Service. It was a beautiful day, and Verena drove me to the station full of high hopes at six-thirty a.m. I got to the office about eleven and was received in a friendly and informal way by the big boss. He told me what is written in this letter to Salam which I enclose. After this interview I walked around Washington for some hours, sweating out a decision. I was still free to go to Russia, and I would stand a very good chance of getting back. If I would go and get back safely, it would be a victory for the international brotherhood of science, showing the world that a courageous individual can place his trust in the good faith of Russian scientists and not be disappointed. On the other hand, if I would go and get into trouble, having my name attached to a political statement condemning the American Imperialists, making me appear to be a Communist Dupe, I would show the world exactly the opposite. In that case I would show the world that an individual placing his trust in the good faith of Russian scientists is an idiot. It all depended on whether I myself trust the Russians not to turn the conference into a political junket.

I decided finally, I trust the Russian scientists, who are genuinely interested in discussing the kind of work I have been doing. But I do not believe these scientists would have any control over the situation, if the local political people decided to interfere. So I was forced to the conclusion, I do not trust these people to keep it nonpolitical; and if I do not trust them myself, it is crazy to risk my job for the principle that one ought to trust them. After I made up my mind about this, I walked over to the National Gallery and relaxed my spirit with Rembrandt and Cézanne. Perhaps if Washington had not been so beautiful that day, I would have decided differently. There was such a peace over the city, with its wide grass spaces where you can walk for miles without meeting anybody, its clean white buildings and brilliant blue sky. It would be very hard to say good-bye to all this.

[ATTACHMENT] MARCH 15, 1955

Dear Salam,

I am sorry to say I had to give up the Moscow conference. I feel very bad about this, but I decided not to risk it. Yesterday I went to the head office of the Immigration Service in Washington. They were intelligent people and on the whole cooperative. They said: (1) They are specifically forbidden by law to give me any guarantee that I may reenter the U.S. (2) If I go to the Moscow conference and get involved in no political manoeuvres, this would not be a reason for them to exclude me from the U.S. (3) If I get involved in any political activities, if my name is used by the Russians for any propaganda and gets into the newspapers, if the conference starts to pass resolutions condemning the American imperialists, or anything of this kind, then I am classified as a "Communist Dupe." (4) The Immigration Service made its own enquiries to find out whether this conference was a bona fide scientific affair or not and was unable to obtain any information. If I rely on my own judgment in believing that the conference is nonpolitical, the risk is entirely my own. If it turns out that I have been duped, I am not to expect any sympathy from the Immigration Service. I myself believe strongly that this is a genuine nonpolitical conference. But the risk is there, and for me the penalty is too heavy to take a chance on it. Best of luck!

> Yours ever,
> Freeman.

APRIL 14, 1955

I knew you would be opposed to my going to Moscow, and I was glad you had the tact to say nothing about it until the decision was made. I understand very well your point of view. However, I know now that there was in fact no political propaganda attached to it. It would have been perfectly all right if I had gone, and I am now sorry I was not brave enough to take a chance on it. The story came out last Wednesday, when the Moscow radio broadcast a statement that the conference had taken place. The statement was entirely factual and nonpolitical in content, and it was clear that the conference itself had been nonpolitical. The physicist Igor Tamm, whose name I know well, made the broadcast, which was sent out on the day the conference ended. There was a list of the people who were there, and he said, "Unfortunately the Americans Feynman and Dyson, who at first accepted our invitations, were prevented by their government from attending." That was all he said about us, and I think it is not objectionable.

When this Moscow radio broadcast was received, the *New York Times* Russian expert called me up and read me the statement, and I told him my side of the story. The result appeared on page one of the *Times* on Thursday. I was glad to see this account also kept strictly to the facts. It is easy to be wise after the event. If I had to make the same decision again without knowing more than I did at the time, I should again say no.

My parents were against my going to Moscow for the same reason that I decided against it. They were afraid that I might be walking into a political trap.

Of the people who were there, two are personal friends of mine. One is Gunnar Källén, a young Swede who was here at the institute last year. Of course Sweden made him no trouble about such a visit. I am very glad he went, because I shall see him before long and hear about it in detail. He will have picked up whatever there was to be picked up. The other friend of mine is Ning Hu who came from Peking. He was a research associate at Cornell in my time, and we used to talk about all kinds of things, especially about China. After the Communist government was established, he decided to go back home. He was aware of the difficulty of living in a Communist society and expected to have plenty of troubles. He had his roots so strongly in China that he had to go back ultimately, and he thought the sooner he went, the easier it would be. After he went back, I heard nothing about him until now. It is good to know that he has established himself and is in good standing. If this were not so, he would not be sent to Moscow.

The purpose of the 1955 Moscow conference was to open up communication between the USSR physicists and the rest of the world. The conference was only partially successful since few visitors came. In 1956 there was a highly successful conference with a much larger number of visitors. Since that time, the Russian physicists have never been isolated from the world community.

I was invited by Charles Kittel to spend a second summer in Berkeley. We rented the same house that we had occupied in 1953, and I worked again happily with Kip and Kittel on problems of solid-state physics. Out of this work grew a theory of spin waves. Spin waves are waves of magnetization running through the atoms of a solid material. I calculated spin waves by the same mathematical method that I had used for quantum electrodynamics five years earlier. This was the same strategy that failed when I tried to use it to calculate nuclear processes at Cornell. To my delight, the strategy succeeded magnificently when I applied it to spin waves.

OCTOBER 21, 1955

I have been working at high pressure from the day we got home. A lot of accumulated jobs connected with the institute and its affairs. To bring in a few extra dollars, I am translating Russian papers for the American Physical Society. Yesterday I did one written by our friend Pontecorvo. He is certainly not the great genius that the newspapers picture him to be. What he does is solid and sound and useful and a bit dull.

Bruno Pontecorvo was a brilliant Italian experimenter who worked as a young student with Fermi in Rome and joined the Communist Party in Paris. As an Italian Jew, he had to escape from occupied Europe. He worked at the Canadian nuclear energy project during the war and at the British project at Harwell after the war. In 1950 he was on holiday in Italy and abruptly disappeared with his wife and two sons. In 1955 he reappeared at the meeting in Moscow which I failed to attend. He was working as an experimenter in the Soviet accelerator project. He spent the rest of his life in Russia as a scientist in good standing, hampered by the Soviet bureaucracy that surrounded him. It never became clear whether he had been a Soviet spy in Canada or in England. He died in 1993. His biography was written by Frank Close and published with the title Half-Life *(2015).*

NOVEMBER 22, 1955

My own work continues to go well. I am beginning to move closer to a field I have for a long time dreamed of working in, the controlled application of thermonuclear (fusion) energy. I shall become a citizen in 1957, and I can start planning to go into secret work about that time. The work itself is getting rapidly less secret, and there is talk of making it completely open. It is no longer necessary to have more than the usual type of clearance to work in it. We are now allowed to mention that there is a big laboratory here in Princeton (connected with the university, not the institute) working on the problem. The director is an astrophysicist called Lyman Spitzer. What they are trying to do is to make a small artificial star. Spitzer is a good man for the job. The problem at the moment is not so much to build gadgets as to explore the basic theory, and I am well qualified to work on it. I believe I shall gradually get more involved in this kind of thing, and I find the prospect exciting. This week I had my first definite offer to work on such problems. It would have been a full-time job in California with a huge salary and an industrial environment. I said no to that. But I will have no difficulty in getting into the work part time without cutting myself off from academic life.

I never worked with Spitzer's fusion project in Princeton. After sixty years the project is still active, but it never achieved its aim of producing fusion energy at a cost competitive with coal and oil. In my opinion, the whole fusion enterprise made a strategic mistake around the year 1960, when it moved too soon from exploratory science to large-scale engineering. After 1960 all the fusion projects were building

*big machines with a few fixed designs, intended to demonstrate economic produc-
tion of fusion power. The big machines failed to be competitive, and there was no
support for scientific experiments trying out radically different designs on a smaller
scale. I was lucky not to become trapped in these fruitless attempts. Instead, I
found more exciting challenges in the field of fission energy.*

12

MOSCOW AND LA JOLLA

THE YEAR 1956 was mostly concerned with new ventures going outside my academic role at Princeton. The first venture was an international meeting of physicists in Moscow. I went to the meeting and spent a few days in Leningrad, fulfilling an ambition that dated back to my days as a schoolboy in England. It was a great joy to get to know the Russians, both as scientific colleagues and as human beings. The many hours and days that I had spent studying the Russian language were richly rewarded. Only by knowing the language can a visitor see below the surface of an alien society. The Russian language is the key to the Russian soul.

My second venture was starting a new career as a nuclear engineer. I spent the summer at La Jolla in California, where a group of scientists was invited to launch a new company with the name General Atomic, building and selling fission reactors in the commercial market. In three months we finished a preliminary design for a reactor called TRIGA, the name meaning Training, Research, Isotope-production, General Atomic. The reactor was a commercial success. The company sold seventy-five of them, mostly to hospitals and medical centers where short-lived isotopes were needed for diagnostic purposes. The TRIGA differed from other reactors because it was designed with safety as the primary consideration. Other reactors relied on engineering for safety. The TRIGA relied only on laws of nature.

My job in Princeton gave me unusual freedom to engage in extraneous activities. I always had a short attention span, jumping frequently from one enterprise to another.

JANUARY 3, 1956

We were lucky to have a long spell of hard frost, unusual for Princeton. Lake Carnegie is frozen over with ice eight inches thick. I bought Katrin a pair of skates and took her out to the lake with me every day. She has learned amazingly fast (the ballet must have helped her with this), and we have got along together famously. One afternoon we skated five miles together from one end of the lake to the other and back again. During this week I found time to finish my spin wave papers. I feel quite light-headed after five months with spin waves chasing each other around in my brain.

JANUARY 25, 1956

I have strong opinions about space travel, and I give them to you for what they are worth. Technically there is no doubt it is possible with existing equipment (given five or ten years for building and testing the machinery) to transport a group of people to the moon with enough supplies to last a year or two, and bring them back alive. The weight of oxygen needed to supply one man for a year is surprisingly small, less than the weight of food he will eat; if he stays longer than a year, he will grow plants which produce both food and oxygen. Meteorites; the risk of being hit is very small, except for the very small particles of dust which can be stopped by a thin roof.

The question how soon people will go to the moon depends on how much money and effort the human race decides to spend on it. It will certainly be expensive (estimates vary between ten and one hundred million pounds) and not immediately profitable. I believe it will go fast because of the psychological situation in Russia and America. Each side is convinced it has to get ahead of the other in such enterprises. It is probably true that to have an observation post on the moon with a fair-size telescope would be a military advantage for the side which gets there first. I believe the job will be done, by both Russia and America, within the next twenty years, perhaps much sooner. And once the first trip is made, there will be no end to it, it will become as commonplace as flying the Atlantic.

This prediction turned out to be half right, half wrong. The last sentence is extravagantly wrong. I wonder now why I was wrong. Perhaps the main reason is that I grew up during World War II, when big enterprises were started and finished regardless of cost. I was still thinking in the wartime way. During the war, money did not matter. In peacetime, the political game was played with different rules. Money mattered.

FEBRUARY 13, 1956

Today I made a speech at the institute and let them have my spin wave theory for the first time. It went well. Oppenheimer and Pauli (from Zürich) were there and seemed to be excited too. On Friday I shall repeat it at Columbia, and later on at Pittsburgh. It is fun to go around with something new to talk about. The last two days I spent digging up information about Hermann Weyl for an obituary notice which will appear in *Nature*. The difficult part was deciding what to leave out, as they gave me a total of only 750 words. I could easily have written three thousand.

Hermann Weyl was the only person at the institute with world-class stature both as a mathematician and as a physicist. He was largely responsible for making symmetry-groups the central concept of modern particle physics. He retired from the institute and returned to Switzerland one year after I joined the faculty. I was lucky to make friends with him during his final year in Princeton. He died one year later in Zürich.

February 29, 1956

I got invited again to Moscow, this time to a conference from May 14 to 20, and this time I intend to go. The arguments which persuaded me not to go last time are entirely absent, now that I know last year's conference to have been a genuine and nonpolitical affair. I still shall take a risk not to be readmitted to the United States when I return, but this I am prepared to accept; in case of trouble my position will be strong enough to fight it through successfully—the whole American scientific public will fight for me. I shall probably leave for Russia on May 12. Already I started taking Russian classes at the university, and these I am enjoying very much. It is fun to be a student again and sit with boys and be scolded for making grammatical mistakes.

The death of Stalin in 1953 started a period of rapid change in Soviet society. Symbolic of the changes was a novel, The Thaw, *by Ilya Ehrenburg (1955). Ehrenburg was a friend of Nikita Khrushchev, who became leader of the USSR in the same year and made the thaw real. Four important consequences of the thaw were: (1) the release of millions of prisoners from the gulag camps, (2) a speech by Khrushchev officially denouncing the abuse of personal power by Stalin, (3) loosening of censorship of speech and writing, and (4) the opening of the USSR to foreign visitors and international meetings. Ehrenburg was a loyal Communist, considered by dissident Russians to be a party hack. His novel carried a powerful message, with dripping icicles and budding leaves of spring presaging the rebirth of human feelings after a long freeze. In 1956 when we visited Russia, the thaw was in full swing and hopes of further liberation were high. In retrospect, we can now see that the thaw was real and permanent, but the hopes for the future were exaggerated. The USSR never went back to Stalinist terror and never went forward to Western-style freedom.*

March 17, 1956, Pittsburgh

There is much talk about Russia. Invitations have been sent out wholesale, not only to our high-energy conference in Moscow but also to a solid-state conference a week later in Sverdlovsk just the other side of the Urals. Everybody who is invited is happy because the others decided to accept. No one is refusing to go. We shall be a big and merry crowd. Amongst others, Peierls is definitely going and also Bethe from Cornell. I have sent in my application for the Russian visa and was happy to find on the application form I must identify myself as Freeman Georgievitch, as in the old Russian novels.

The American Atomic Energy Commission has completely reversed itself since last year when they stopped Feynman from going to Moscow. I am at the moment under investigation because I am applying for AEC clearance to do secret work during the summer in California. I wrote a letter, not asking the AEC for permission to go to Moscow, but informing them that I shall go and that I shall understand it if in consequence they do not want to give the clearance. The reply came back by telephone that the clearance will not be affected one way or the other. The AEC has also given its blessing to Bethe who knows much more about hydrogen bomb secrets than Feynman ever did. It seems some common sense has finally penetrated into the AEC.

Just before I left Princeton, there came a telephone call from the AEC asking, "Who are the Princeton people who are going to the Moscow conference?" I thought, "Well now, here starts the trouble." Then the AEC official said, "Please tell them that if any of them are working on AEC supported contracts, we shall help pay their fares to Russia." It looks as if this opening of communications will be more or less permanent, and there will be many more trips to Russia in years to come. This makes me happy that I did not lose much by being cautious and refusing to go last year.

APRIL 4, 1956, ROCHESTER, NEW YORK

Today I had the pleasure of putting to good use the Russian I have been studying for so many years. I had supper with Veksler the Russian accelerator expert. After a while I found myself saying *"Nyet spasibo"* to the American waitress, which caused some amusement. Veksler said they are all so much looking forward to having me in Moscow that I must be careful not to be questioned to the point of exhaustion by eager young theoreticians. It is lucky that we got this chance to make friends and overcome the language problems before the trip to Moscow. I shall feel no strangeness at all when I go there. Veksler said I was welcome to come to lecture in Moscow for a few months "whenever you find it convenient." It is all somehow intoxicating.

MAY 15, 1956, HOTEL MOSKVA, MOSCOW

I am safely here and settled in. I feel as if I had already been here a long time. The Russians work terribly long hours. We have meetings every day from ten till two and from five till nine. For each session we drive half an hour in a bus there and back, so not much of the day is left. I am amazed to find what an exaggerated reputation I have in Russia. This is partly because I am the one who reads and reviews their papers. They treat me with enormous respect. And they have evidently read what I write, not only in the *Physical Review* but also in the *Scientific American* and even the obituary of Weyl in *Nature*. The meetings themselves, apart from being too long, are lively and interesting. The Russians are informal, more like Americans than Western Europeans, and they contradict and argue and make jokes freely in the meetings. Last night I managed to get out for an hour and walk around the streets. It is a little like Paris. Lots of cheerful people strolling around and enjoying the warm evening, chattering and laughing. Most of the houses in a state of genteel decay with plaster and paint flaking off. Many cafés open with tables and chairs on the sidewalk and people drinking wine or coffee. How I should love to be here for two weeks with nothing to do but sit around and explore the city. But we shall be organised almost every hour of our stay. After two weeks I shall be ready to come home.

The Russia that I saw in 1956 and on several later visits was no Potemkin village. I knew the language well enough to see the fears and tensions below the surface of the society. Russians talked to me with astonishing freedom about the corruption and incompetence of their government. The thaw was real, and the ice that held Stalin's Russia in place was softening. Russia was already moving, in a slow but massive slide, towards the peaceful collapse of the USSR that took us all by surprise thirty-five years later.

In 1955 there was an international conference in Geneva, at which government officials from countries with nuclear projects met to decide how much of their secret information should be made public. Until that meeting, all of nuclear science and engineering was secret. It clearly made no sense to keep secret the basic facts about peaceful uses of fission and fusion energy. Delegates from the United States, the Soviet Union, Britain, France, and Canada agreed to share and publish nuclear information that was not related to bombs. One of the American delegates at Geneva was Freddy de Hoffmann, a young scientist from Los Alamos. He understood that the Geneva agreement opened up the field of peaceful nuclear energy to private business. After the meeting, he resigned his position at Los Alamos and founded a nuclear start-up company which he called General Atomic. He raised enough money to rent a schoolhouse in San Diego for the summer of 1956 and to invite a big group of experts to sit in the schoolhouse for three months and decide what the new company should do. The group had roughly equal numbers of physicists, chemists, and engineers. I was one of the physicists. I arrived at the schoolhouse in June 1956 knowing nothing about nuclear reactors but eager to learn. I was housed with the other visitors in a big motel in La Jolla.

My first impressions of La Jolla were unfavorable. It seemed to be a good place to take a holiday but a bad place to live. Abundant flowers that were actually real but looked artificial beause they were too brightly colored. Houses with big swimming pools but no books and no reading lights. Streets full of big fast cars with few pedestrians. After we lived in La Jolla with our family for a few months, we began to like it better. There was a good public library where our children could borrow unlimited supplies of books. There was a good local hospital where one of our babies was born. The Scripps Institution of Oceanography was a world-class research center where we made lasting friendships. The famous Doctor Seuss,

author of children's books that our children loved, came in person to spend a morn-
ing with our son's first-grade class, collecting ideas from the class for his next book.
In the end we liked La Jolla so much that we bought an apartment there.

WEDNESDAY, JUNE 20, 1956, LA JOLLA

The work may turn out to be very exciting, or it may be a terrible flop. It is hard to say at present. In many ways it reminds me of
Bomber Command. A group of us has been given the job of thinking
up a nuclear reactor which shall be absolutely safe, so it can be played
around with by untrained people and there can be no question of it
blowing up. Such a reactor would be greatly in demand for hospitals
and such places where they need a reactor but do not want to maintain
a staff of physicists to take care of it. This is a clear enough assign-
ment, and if we can do something along these lines, it will be exciting.
On the other hand we have the feeling, as we did at Bomber Command, that we are remote from real life. Few of us know anything
about the practical construction of reactors, and we do not have any
experimental facilities here. So all we can do is to think up general
ideas and follow them to a preliminary design stage. Perhaps something good will come out of it.

Edward Teller was a famous Hungarian physicist who came to the United
States in 1935 and became passionately involved in both the military and civilian
applications of nuclear energy. He had two obsessions, the hydrogen bomb and
civilian reactor safety. The hydrogen bomb obsession was well known to the public
and made him generally unpopular. The reactor safety obsession was not so well
known but equally important to Teller himself. He was chairman of the Reactor
Safeguards Committee of the U.S. Atomic Energy Commission for many years.
He was a friend of Freddy de Hoffmann at Los Alamos and was an enthusiastic
supporter of Freddy's venture into civilian nuclear technology. He came to Gen-
eral Atomic in 1956 because he saw it as an opportunity to build reactors far safer
than any existing reactors. He believed passionately that nuclear energy could be
a blessing to mankind if reactors were perceived by everybody as safe. It was not

enough to solve the technical problems of safety. If nuclear energy were to be acceptable in the long run, it was necessary to make it visibly and obviously safe, so that we could operate a reactor in the middle of a city and insurance companies would not hesitate to insure it. When the visitors assembled at General Atomic, Edward Teller immediately took charge of the group working on the design of a safe reactor. He announced that our goal was to design a reactor so safe that we could hand it over to a bunch of high school kids to play with and they could not get hurt.

JULY 15, 1956

I am more deeply involved in nuclear reactors than I would have believed possible four weeks ago. There is lots to be done, and this is a good place to do it. Our group which is working on safe reactors has done so well that people are now being transferred into it from the two other groups (test reactors and ship reactors). I was responsible for most of the new ideas which gave us our lead. The rest came from Edward Teller who is also in our group. It is exciting and infuriating to work with Teller. I had often heard about scientists behaving like prima donnas, and now I know what it means. We had yesterday a long meeting at which I disagreed with him, and he was in a filthy temper. Finally he won the argument by threatening to leave the place if we would not do things his way. I did not know whether to laugh or cry, but it was clear that the best thing was to laugh and go along with him. I do not have to take this seriously. But I understand now what a misery he must have been for Oppenheimer at Los Alamos. Oppenheimer could not let him run the whole show his own way. I am glad I am not likely ever to be Teller's boss.

We are trying to design reactors which shall be intrinsically safe against blowing up or melting down, so that one can put them in hospitals or factories, and the neighbors will not be afraid of them. Our quarrels arise because I have a scheme which will take two or three years to develop and is very good, while Teller has a scheme which can be done with luck in one year but is not so good. Teller says, since this is a small company, we ought to drop everything else and work on his scheme to get it done as quickly as possible. We decided yesterday to do this. I am not particularly unhappy about it. I shall enjoy working on his scheme. But I am glad I put up a fight and made myself heard. It is astonishing how quickly one can become the world's greatest expert on safe reactors. It is like Papa and his handbook of grenade fighting. I have never seen the inside of a reactor, and I did not even read about one till four weeks ago. Teller is in roughly the same situation. What makes me happy is that I have now an established reputation as a reactor expert, whether my ideas are ever adopted or not. I shall have to do nothing more, and invitations to consult with various companies will come flowing in. And I find this reactor business genuinely exciting. As time goes on, I begin to like this place better. Today I shall go down to Mexico and see my first bullfight.

My father was a professional musician who volunteered to be a soldier when World War I broke out in 1914. He was quickly appointed grenade officer for his infantry unit, since nobody else wanted the job. He wrote the official handbook of grenade fighting (1915), never having seen a real grenade. His handbook was used by the entire British army thoughout the war, and also by the American army when the United States joined the war in 1917. He was thirty-one when he became the expert on grenade fighting, and I was aged thirty-two when I became the expert on safe reactors. All it takes to become an expert is a little common sense and a little imagination.

AUGUST 12, 1956

I am having a joyful time here. I seem to have made quite a dent in the atomic energy business and their security regulations. What I have done for this company seems to have impressed them. They are begging me to stay here and be the head of the theoretical division. They offer to do absolutely anything to make it agreeable to me. Luckily we had here for the last week a young man called Lewis Strauss, Jr., who happens to be the son of the Lewis Strauss who is the head of the whole AEC. Strauss Junior is a physicist, and I found him congenial and easy to talk to. The old Strauss is a firm Republican and would decide the question in favor of General Atomic if it ever came to his notice. Yesterday before he left to go back to Washington, young Strauss whispered in my ear, "We will have you cleared inside two weeks." So I imagine the old-fashioned power of family influence will be brought into action on my side. Meanwhile I am amusing myself with reactors, and I find it absorbingly interesting to think about them. Probably this summer is a turning point in my life. I find the atomic energy business congenial, and also I am good at it. My real talent is perhaps not so much in pure science as in practical development. Just as Papa would never make music to himself in an ivory tower but always in the context of a particular group of people who would play it.

Sunday, August 26, 1956, La Jolla

You may like to hear the sequel to my quarrel with the AEC. A few days after my previous letter to you, the news came from Washington that I am cleared. But I am only cleared for one project, which is the most secret of all the things General Atomic is doing. The AEC has a logical explanation for this absurd situation. Their regulations say that secret information may be given to a foreigner only when this is necessary to the national defense. Obviously, if the information is not vital military information, it cannot be necessary to the national defense to give it to me. Therefore I can have the important military secrets but not the unimportant civilian secrets. It is the craziest joke I ever heard. A few days after the clearance, there came a telephone call from Los Alamos asking me to sign on with them as a consultant and to stay a week on my way home from here in September. The contract allows me to go to Los Alamos anytime I like. I have agreed to sign on.

September 20, 1956,
the Lodge, Los Alamos, New Mexico

I finally managed to get here. Yesterday came the news that my clearance has been approved. I flew to Albuquerque in the evening and this morning took the little five-seater plane which comes up here. The trip from Albuquerque to here is marvelous, mountains and deserts glittering in the morning sun. The plane lands on a little mesa not much wider than the runway, perfectly flat on top and with a deep canyon on each side. From up above Los Alamos looks like a village, tiny among these vast distances. Most conveniently, I took the last of my fourteen rabies injections in La Jolla yesterday. I got through that very well. Now I have two days here, then on Saturday I fly east. Today I spent absorbing all the information I could at tremendous speed. My clearance is good for everything, all kinds of bombs included. I have done pretty well learning all this stuff in one day.

The rabies injections were needed because I had been bitten by a stray dog in Mexico. In those days the injections were made from rabbit brains containing antibodies to the virus, so there was a risk that the brain tissue might cause an encephalitis as disabling as rabies. Edward Teller helped me to evaluate the risks of encephalitis and rabies with and without the injections. He advised me to take the injections.

I never needed to worry about all the bomb secrets that I was carrying around in my head. Fortunately, there was always a clear separation between the secret stuff and the open science and the political activities that I was free to talk about. I did not worry whether my letters from Los Alamos to my parents might be read by some snooping security people. I did not know whether they were snooping, and I did not care. They had the right to snoop if they wished. There was never anything in the letters that came close to being secret.

After Edward Teller and I and the other summer visitors departed from La Jolla, the people who remained at General Atomic reversed the decision we had made in July. They decided to go ahead with the safe reactor project using my design instead of Teller's. The main difference between the two designs was that Teller used engineering tricks to make the reactor safe while I used laws of nature. A reactor emerged within a year from my ideas. The company designed it, built it, licensed it, and sold it, all within two years, a time scale that would be unthinkable today. They sold altogether seventy-five reactors, mostly to hospitals. None of them gave any trouble to their owners, and several are still running now after more than half a century. The person who did the detailed design of the reactor was Massoud Simnad, a brilliant Iranian chemist, settled in the United States but enjoying friendly relations with the Ayatollah Khomeini. Nuclear engineering requires more chemistry than physics. The physical ideas that I contributed were simple, while the chemical ideas that Massoud Simnad supplied were complicated and brilliant. The essential trick that made the reactor safe was to load the fuel heavily with hydrogen in a chemical form that remained stable at high temperatures.

NOVEMBER 13, 1956

Last Tuesday we sat up with the television at the house of some friends, watching the election results come in. This is a ritual which has a dramatic quality, even when the results are known in advance. In this case the presidential vote was no surprise, but there were many exciting struggles in the Senate and House, and these mostly ended satisfactorily with a Democratic majority. At one-thirty Stevenson came to the camera and made his speech conceding the election, then Eisenhower came and made his speech, and we went home to bed. Eisenhower was impressive and Stevenson not. I guess Stevenson is fed up with the whole business.

A lot of talk about the affairs of Hungary and Egypt. People here are mostly strong Israel supporters and were happy when England and France attacked Egypt. But it seems this view is not widely supported in the rest of the country outside Princeton and New York City. One hears very often the statement that Nasser is like Hitler and it is necessary not to repeat the mistake made in 1936 when Hitler reoccupied the Rhineland without interference from France. My opinion is that this statement is correct and still offers no solution to the problem. I believe if France had invaded Germany in 1936 to remove Hitler from the government, the whole world (especially America) would have held up their hands in horror. And the end would have been a French retreat and German victory, just as it was in 1924. It seems clear now that the policy of attacking Egypt will not be maintained firmly or consistently enough to do any good.

In 1923 the German government refused to pay the reparations imposed by the Versailles Treaty, and France sent an army to occupy the Ruhr area of Germany and enforce the Treaty. After a while, the German economy collapsed and the French army of occupation withdrew. The main result of the invasion was to strengthen the popular support for Hitler in Germany.

I heard some interesting remarks about the Hungarian affair from my neighbour George Kennan. He said, from his whole experience of Russia, the behaviour of Russia during the first week of the rebellion could not be explained except by some paralysing crisis in the Russian government. He suggests as the most likely explanation that the Red Army refused to take on the job of policing the satellite countries indefinitely. Then when the rebels went too far in their anti-Russian policy, even the army agreed that they had to put a stop to it.

The Soviet army occupying Hungary stood aside and did nothing for a whole week, while a group of anti-Communist Hungarian rebels took over the government. The delay of the Soviet response astonished Kennan. At the end of that week, the Soviet troops finally moved to crush the rebellion. The rebels mostly escaped from Hungary to Austria and found refuge in Western Europe and America.

DECEMBER 14, 1956

Last weekend we were brave enough to drive up to Ithaca. The children loved to walk on the frozen lake (we have no real winter yet here) and throw snowballs at each other. We stayed with some people called Gold whom I knew well in Cambridge ten years ago. Tommy Gold is a brilliant fellow and is equally at home in physics and astronomy. He got the job of chief assistant to the Astronomer Royal when I refused it four years ago. He was a great success there and got along

well with [Harold] Spencer-Jones. Spencer-Jones understood that his
own job was to run the observatory and allow the younger people a
free hand to research and innovate. Tommy instituted a cosmic ray
program at Herstmonceux and was by far the best man they had there.
Then came the retirement of Spencer-Jones, and [Richard] Woolley
was put into his place. He had other ideas. One day he asked Tommy
what he was doing, and Tommy said he was preparing a paper for a
conference in Stockholm about the Sun's corona. Woolley said, "But I
never gave you permission to work on the corona." That was enough
for Tommy. He resigned then and there and accepted an invitation
from Cornell. Now he has been offered a good permanent chair at the
Harvard Observatory, and so he will stay in this country. Woolley
seems to be a pathological character. Now that Tommy has left, his
friends in the cosmic ray group at Herstmonceux are not allowed to
correspond with him. All communications must go through Woolley
himself. Tommy showed me the letter from Woolley in which this
decision was announced, written in the finest civil service circum-
locutions. Really it is a pity this has happened. There was a chance
something might have been done for the Greenwich Observatory, but
now it is hopeless.

*It was lucky for me that I had the good sense to refuse the job at Herstmonceux
in 1952. If I had accepted, I would have ended up as Tommy Gold did, in a bruis-
ing fight with Woolley.*

It is good to hear that you in England are making a big effort
for the Hungarian refugees. Here there is a great organization for
bringing them in and finding them homes and jobs. Luckily there is
a substantial population of Hungarian-speaking people already here,
and that makes it a lot easier. So far the government agreed to take in
21,500, which is all they can do until Congress is in session. Probably
Congress will be happy to raise the number if more want to come
here. The only objection I have heard to this is from some physicists

who are nervous about the effect of having 21,500 Edward Tellers in the country.

One of the first things I did when I became an institute professor was to push for the appointment of Bengt Strömgren, a world leader in astronomy. I wanted our School of Natural Sciences to include astronomy as well as physics. Oppenheimer agreed to invite him, and he came in 1957. He was outstanding as an observer as well as a theorist. In his office at the institute he had a little machine of his own design, a personal computer before personal computers existed. He observed A stars in the sky and measured their ages by accurate measurements of brightness in four colors. A stars are bright stars that are easy to measure accurately. He put the data for each star into his machine. The machine then calculated its orbit around the center of the galaxy and deduced the place where it was born. After he had plotted the birthplaces of a few hundred stars, he could see that the births at any time fell into a spiral pattern, and the spiral moved around the galaxy as the time of birth advanced. So the births of the A stars revealed the past history of the spiral arms of the galaxy. This was the most elegant piece of observational astronomy that I ever saw. It was all done on Strömgren's desk top with a machine costing a few hundred dollars.

A pleasant thing happened this week. For some years I had had the idea that it would be interesting to look for variable white dwarf stars. A white dwarf is so small and dense that its natural vibration period would be a few seconds, instead of the hours or days which ordinary stars have for periods. So if there is a variable white dwarf, it would not be recognised by the usual method of taking photographs with an exposure time of several minutes. The other day I mentioned this idea to Strömgren who is our leading astronomer, and asked him if anyone had ever looked at a white dwarf with a light detector having a rapid response. He said there is one white dwarf called o2 Eridani B which is most suitable for this experiment. It is reasonably bright (ninth magnitude) and not close to another brighter star. It happens also to be in the part of the sky (just below Orion) which can best be observed in January. So this is the right time to do it. The chances of finding pulsations in this particular star are quite small. Still it is worth a try. Kitt Peak is a brand-new observatory on a mountain in a remote part of Arizona. It was chosen for the National Observatory because it has better seeing than California. So next January 4, think of o2 Eridani B, and hope for clear skies in Arizona.

The skies were clear, and Strömgren did the observation. o2 Eridani B did not pulsate.

13

THE FORSAKEN MERMAN

"*THE FORSAKEN MERMAN*," *written by Matthew Arnold in 1849, was one of my mother's favorite poems. Arnold wrote it as a reaction to the Hans Andersen story, "The Little Mermaid," which was published in 1837 and later made even more famous by Walt Disney. Andersen wrote a wonderfully sentimental tale of the girl who abandons her home in the sea to live with her prince on land. Arnold is telling the same story from the point of view of the husband and children that she abandoned.*

> *Children dear, was it yesterday*
> *(Call yet once) that she went away?*
> *She will start from her slumber*
> *When gusts shake the door;*
> *She will hear the winds howling,*
> *Will hear the waves roar,*
> *Singing, here came a mortal,*
> *But faithless was she,*
> *And alone dwell for ever*
> *The kings of the sea.*

It is a long poem, with the refrain "Children dear, was it yesterday?" repeated at the beginning of each verse. My mother knew it by heart and used to recite it to my sister Alice and me when we were little and she put us to bed. It is not a great poem, but it was exactly appropriate to my situation in February 1957, when my wife Verena walked out of our home and abandoned her husband and children. In the New Year letter one month earlier, there is no hint of impending drama.

NEW YEAR'S DAY, 1957

I shall be five weeks away from home before the Columbia term begins. First a spell with General Atomic at La Jolla, then at Los Alamos, and finally ten days at Aspen, Colorado. For the last ten days Verena will fly out to meet me, and we shall take a real holiday together.

We had Hans Haefeli to stay over Christmas, and he as usual made up for the aunts and uncles and godfathers which we lack in this new country. A fine and faithful friend he is to all of us.

At the moment my main reading matter is the history of the Americans in Petrograd in 1917–18 by George Kennan [1965], who occupies the next office to mine at the institute. This book is delightful reading, it is history written according to the maxim *Le Bon Dieu aime bien les Détails*. I am reading volume one while Kennan is getting volume two ready for printing.

Kennan became an institute professor in 1956. His two-volume work was published with the title Soviet-American Relations 1917–1920. *He remained a friend and colleague until his death at age 101 in 2005. My most vivid memory of Kennan when he came to the institute is his story of the Trotsky papers. In the State Department archive in Washington, he discovered an old shoebox containing a hundred handwritten documents. Before Kennan found them, nobody knew that they existed. They were the first official communications from the newborn Soviet government to the American diplomats in Petrograd. They were written personally by Leon Trotsky. To discover such a treasure is the mark of a first-rate historian. Kennan and I became friends because we were both passionately engaged in the problems of war and peace and nuclear weaponry. He knew more about history, and I knew more about weapons, so we both had something to learn.*

JANUARY 31, 1957, ASPEN, COLORADO

Now I have a peaceful hour to write you our news before going to bed. Such a beautiful day it was. In the morning brilliant sunshine, dark blue sky, virgin white snow, equally perfect for skiing or for gazing. Verena and I both went up to the top of the mountain and skied down the four-thousand-foot run to the bottom. Verena has made remarkable progress with skiing in only six days here, never having done it before. She has a natural balance and suppleness, and I hope she will have a chance to do much more in the future. We were here just a week, and we go home the day after tomorrow. It was the best, the most complete, the most triumphant holiday I have ever had. Pain and joy, love and bewilderment and laughter, all on top of each other. And in the end, tears and confusion are past, and there remains a shared courage, an understanding, and a lightness of spirit. Now let me explain briefly what has happened. An old friend of mine from Cambridge days has been at the institute for the last year and a half. While I was away in La Jolla and Los Alamos, he and Verena have fallen in love and decided to run away together. Verena came to Aspen to tell me this, and to make a harmonious and dignified end to our marriage.

Georg Kreisel had become one of the leading experts on mathematical logic during the ten years since I had known him as a student in Cambridge. In 1955 he applied for a membership at the Institute for Advanced Study to work with Kurt Gödel. I wrote an enthusiastic letter of recommendation for him, and he was accepted. We remained friends during his stay at the institute until the end of 1956. The drama that followed came as a total surprise. In my letters I did not mention his name until the affair became public.

The practical arrangements are simple. I shall keep the children. We have the competent Imme Jung who will take care of them for the time being. Verena will be in Princeton till about April and will be with us to see the new household running smoothly. Verena will not marry this fellow but will eventually find herself a job somewhere and live in the independence which she has always wanted. I will obtain a divorce as soon as possible, and I will not under any circumstances ask her to come back to me as a wife. I hope she will come back frequently as a friend and as a mother. Later I will write in more detail about our plans for the future.

Just now I want only to make some general remarks about the past. First, about myself. Please do not offer me your sympathy or your pity. I have been happy in this marriage, and I have no regrets now it is over. It has enriched my life in many ways, and this enrichment is permanent. Second, about Verena. You can blame her for what she has done. But I do not. I consider that she has fulfilled her obligation to me, by bearing me two fine children, by caring for all of us through the difficult years when the babies were small and money was short, and by loving me faithfully for seven years. She leaves me now just when our family life is getting to be easy and comfortable, the children soon to be all at school, the finances ample, and a beautiful house to live in. What she has done may be crazy, but it is not irresponsible. I believe that she has earned her freedom, that she is doing the right thing in following her own star wherever it leads. I want you to give her your respect if not your approval. As Blake [1799] says in these lines which I have long known but never rightly understood till now,

> He who binds to himself a joy Does the winged life destroy.
> But he who kisses the joy as it flies Lives in eternity's sun-rise.

FEBRUARY 7, 1957

Today I gave my first lecture at Columbia. This is very good fun. I am going up to New York four days a week, and I have a class of twenty-five students who seem interested and intelligent. It is good for me to get back to teaching, and I find even the train rides in the morning and evening peaceful and refreshing. I leave each morning at eight-thirty and return at six-thirty, like a regular bank clerk. Did you hear in your newspapers about the exciting things that have happened this month at Columbia? This makes it especially worthwhile to be at Columbia just now. As a result of a suggestion made last summer by my friend Frank Yang (of the institute) and another young Chinese called Tsung-Dao Lee (of Columbia), the people at Columbia undertook two novel and clever experiments. The leader of the experimental work was a third Chinese called Chien Shiung Wu who is also exceedingly good. The idea of the experiments is to test the possibility that a spinning particle may carry a definite distinction between its North and South Poles. That is to say, suppose you have a particle spinning in a definite sense (say clockwise) around a north-south axis, and the particle emits an electron, is it possible that the electron will come out with different probability along the north and the south directions? All the theories of the last thirty years assumed that this was impossible. Until Yang and Lee, everybody thought that so obvious that it did not need to be tested. Now the Wu experiment shows an enormous difference between North and South, so big that it could not have been missed by anybody who had looked for it. This is an important breakthrough, and we are all happy about it. Yang and Lee deserve this triumph, as they have been working and thinking about these problems harder than anybody else for the last two years. We have now the job of changing our theories to agree with the new information, and this is likely to lead to substantial progress. It is fine for me to be at Columbia where the experiments are being done, and to have lunch with Lee and Wu and the others whenever I feel like it.

The more I see of Imme, the more impressed I am with her firm and solid character. Without this, of course, any kind of peaceful changeover would have been impossible. Imme is twenty years old and comes from Berlin, so she is used to a life of unpredictable ups and downs. I hope she will be happy with us and will stay her two years.

Imme Jung came to our home as an au pair to help care for the children. Verena had met with her in Austria in summer 1956 and offered her the job. Imme obtained a visa that allowed her to work in America for two years before returning to Germany. When she arrived in New York, Kreisel came with Verena to meet her at the boat. Imme assumed that this gentleman who came to the boat was Freeman Dyson. I did not meet her until I returned home from Aspen in February 1957, two weeks after her arrival in America. When she arrived, she had no warning that she was walking into a family crisis.

Sunday night, February 11, 1957

Just came home from singing the *St. John Passion* of Bach. What a wonderful work it is! Such a joy it was to sing for three hours and forget all problems and worries.

After I came home I went to sit on Katrin's bed for a while, and she had a good cry. Do you remember how as children Alice and I used to love to listen to you reading "The Forsaken Merman"? I have forgotten everything except the refrain "Children dear, was it yesterday?" Yesterday in fact it was, when Verena moved out. Today she came back for the afternoon while Imme had her free time, and she gave the children their supper while I was at the singing. But she is now definitely and officially moved out. On the whole we have come through this crisis very well. Yesterday was the worst day because we had to tell the children. It is worst for Katrin, but all three of them have been brave and cheerful. This makes it easy for me to be brave and cheerful too.

The Oppenheimers gave a big party last night, and Verena and I went to it as our last public appearance together. I shall always be sorry we did not put on as good a show for our wedding as we did for our separation. During the party I took the opportunity to speak to Oppenheimer and told him it was good-bye for us. This was the first time we told anybody here outside the family. Afterwards when we were leaving, Mrs. Oppenheimer came with us to the door and kissed Verena, and tears were streaming down her face. Before I had always found Mrs. Oppenheimer tiresome, but I am grateful to her for those tears. After the party we went our separate ways home.

FEBRUARY 14, 1957

I feel like a man who has been condemned to death and led before the firing squad, has heard the order to fire and the report of the guns, and then discovered that all the cartridges were blanks and that he is still alive and well. He walks away from the scene of execution with a light heart, the earth is still beautiful, and the problems of life can never seem so frightening again. So much for myself. You are right, it is the children who suffer most in this business. But one thing you must remember, in our family the children always knew that Verena had walked out of her previous marriage. So this was accepted by them as an event in the normal run of events, of course not a desirable event, but still not unprecedented and not upsetting completely their feelings of stability. I think this was important in making it easier for them to accept what has happened.

If you would see Esthi and George now as I see them every evening when I come home from my day in New York, you would not believe they had a care in the world. They are their usual noisy lively selves, shouting and squabbling and telling stories without beginning and without end. The only change in their behavior is that they are a little more affectionate to me and give me some of the goodnight hugs and kisses that used to be reserved for Verena. I never had any choice about what to do, nor any difficult decision to make. All I had to do was to take what came, and be brave and cheerful, for the sake of the children. With the children around to help, this was never difficult.

FEBRUARY 18, 1957

For heaven's sake don't think of our story as a tragedy. I much prefer the attitude of Mila Gibbons, the dancing teacher who is now taking care of Katrin. I went and had a long talk, both with Katrin and afterwards with Mila, on Saturday night. Mila told me how it happened that she invited Katrin to stay. Katrin came to her class on Monday a week ago in tears. Mila's daughter Eve found out what was the matter and went to tell Mila. "Oh, only that," said Mila. "I thought it was a death in the family, or a dog run over by a car."

Mila and I see very much eye to eye about the situation as it concerns Katrin, and I will not be surprised if Katrin stays with her for some months. Mila herself ran away from her husband many years ago and has brought up her two children single-handed.

On my train rides from New York in the evenings I have been reading Ibsen's *Doll's House* [1879], which I had not read for many years. It is a grand story, and I think it explains why the word *betrayal* is quite wrong to describe what happened with me and Verena. What makes our situation different from the usual one is that the roles of Torvald and Nora are somehow mixed up, so that at the critical scene in the third act she is Torvald and I am Nora.

Like thousands of other frustrated housewives before and since, Verena saw Ibsen's Nora as a role model for herself. Nora was the first heroine in world literature who had the courage to abandon her husband and children and escape from the doll's house in which she was imprisoned.

MARCH 5, 1957,
BROOKHAVEN NATIONAL LAB, LONG ISLAND

One of the many selfish advantages I have from kicking Verena out of the house is that now Esthi and George come to my bed each morning for half an hour before we get up. They come into my bed very quietly, one on each side, and either go to sleep there or start talking about all kinds of things. George now is passing through the first theological-philosophical stage. Yesterday he surprised me by saying, "You know there are two Gods." I said, "What are their names?" He said "One is called Jesus and he makes people, and the other one is called Bacchus and he makes wine."

Verena found herself a job. I am happy and relieved about this.

On Monday she went down to Philadelphia to be interviewed by the Remington Rand company, a big firm manufacturing and designing electronic computers for all kinds of mathematical applications. The same evening came a telegram with a firm offer of a job, at nine-tenths of the salary on which I was supporting the entire family during our years at Cornell. I do not know whether she will take this job, it is not exactly her cup of tea. But with this telegram in her pocket she can feel safe in bargaining for good conditions with any other outfit she chooses to work for. I am relieved to find she is in demand and can take care of herself. How can it be that I was so happy for seven years to be married to Verena, and now I am so happy not to be married to her?

Thursday morning. Verena said yes to Remington Rand and will start work in Philadelphia on Monday.

MARCH 12, 1957

On Sunday I took Imme with me to sing Haydn's *Theresienmesse*. This was a great success. Imme loves music and had sung this work before, so she had no difficulties. I was happy to find something she can do with me, so that I can introduce her to people.

MARCH 27, 1957

You ask about Verena's work and about her new boyfriend. Both seem to be working out well. Her boyfriend is still in Princeton and still preserving a careful anonymity. Verena is determined to keep his existence a secret. I think this is foolish, but I am respecting her wishes in this matter, and I am glad you are doing the same. The desire for secrecy originates not with her but with him. I felt from the beginning this was a mistake. If he wanted to make her happy, he should have taken her off with a flourish of trumpets to Casablanca or some such place completely away from the children. But he wanted to have her love without the inconvenience and responsibility of acknowledging her publicly. So she has to suffer the misery of being separated from the children and still within reach. For this I despise him and consider him unworthy of her. I do not know how things are with them now, nor do I care. She is still living with him and will continue to do so for some time, but I think it is rather a question of *faute de mieux*. He will go back to England in August, and she intends to move to Switzerland sometime later in the year. After that they can see as much or as little of each other as they please.

Now you can see how suddenly struck I was when I read the last act of *A Doll's House*. Nora is sitting impatiently waiting for the "wonderful thing" to happen, that Torvald should take the blame for her forgery of her father's name. For me, the "wonderful thing" was that Verena should be carried away to Casablanca in a cloud of romance and glory. Then the dénouement comes when the wonderful thing does not happen, and Nora finds out she has been worshipping a tin god for ten years. Then she says good-bye and walks out. And that is exactly what happened to me. Only I took the house and kids along too. You must not think, because I make jokes about Verena now, that I am not grateful to her for all the fine things she did. It is just a relief not to keep up this solemn pretence of infallibility anymore.

APRIL 2, 1957

The night after George's fourth birthday party was the institute spring dance. This is a great event which happens once a year, and everybody dresses up in black ties and evening dresses. I asked Imme to come as my partner, and she got a university student to baby-sit for us so we could dance through the night. It happened that this year it was Oppenheimer's idea to have a dinner party of the institute professors and their wives before the dance, and Kitty Oppenheimer insisted that Imme should come to the dinner too. So Esther and I went into town to buy her a gardenia for her dress, and at seven-thirty Imme and I solemnly entered that majestic dining room with Van Gogh originals around the walls, to sit among the distinguished company of the learned. I felt like Higgins taking his Eliza to the ball; after all, Imme is extremely young and has little formal education. But she carried it off perfectly and seemed quite at her ease with everybody. Oppy especially went out of his way to be friendly to her.

Kitty Oppenheimer was not trying to be a matchmaker. She had already visited Imme at our home to welcome her to the institute community. She had the good sense to see that Imme was alone with my children in a strange country and needed some grown-up friends. The dinner party was a great opportunity for Imme to get to know some of the famous institute people.

After the dinner we went down to the institute where the dance was beginning. Very soon the whole mob arrived, about three hundred people, and there was a good band and lots of young people enjoying themselves. We danced from ten till four with an interval for supper at midnight, and neither of us seemed to get tired in the least. Imme had her fair share of attention from the other gentlemen who are friends of ours. She is light and graceful and could make a good dancer. I think I never in my life enjoyed a dance so light-heartedly as this one. Only one thing she said: "I wish there were some way to make all these people understand that I am your maid and not your mistress." This is a real problem, and I see no solution to it.

Easter Sunday, April 21, 1957

You ask me about Imme. Her father is a country doctor in a little town in West Germany, he works hard and makes a good living. Her mother came from a wealthy Berlin family which owned some kind of factory and lost everything during and after the war. Imme's upbringing was impoverished gentility, combined with the chaos of war and Russian occupation. She had a year in England, as a guest in an English family. After that a year in Spain where she was governess to a family of nine kids ranging in age from one to fifteen. In English and Spanish she is completely fluent. We never talk anything but English.

After sixty years I can see more clearly how lucky we all were to have Imme arrive at our home when she was needed. Verena found Imme herself and understood what a treasure she had found. Verena was always grateful to Imme for holding the family together when she walked out. For me, being a single dad was an ideal situation. Having Imme as a young and capable helper made the situation even better. I had always been better as a father than as a husband. When I first met Verena, it was Katrin who brought us together, and being father to Katrin was as important as being husband to Verena. When Verena left, Imme became to me like a grown-up daughter, a delightful companion without the complications of being married. We told the children that Imme would leave and go back to Germany after two years. They said later that they always hoped and suspected that Imme would stay. My parents had found Verena difficult and often hostile when I was married to her. They were secretly glad when Verena left and immediately gave a warm welcome to Imme. During my two years as a single dad, I held back from any display of affection for Imme, but it was obvious to us all that she would fit well into the role of wife to me and mother to my children.

APRIL 29, 1957

Did you see the comet? Tonight at nine o'clock it was a beautiful sight, with a clear moonless night to show it at its best. I pulled the children out of their beds and took them out in the garden with the telescope to look at it. We also saw three moons of Jupiter. The comet was magnificent in the telescope. I am happy the children could see it, and they will long remember it as something special.

On Friday I rushed down to Washington to be chairman of a meeting of the American Physical Society. This was exciting because there was announced the result of an experiment at Columbia for which I worked out the theory. The experiment has succeeded beautifully and gives a result exactly opposite to what everybody expected. So I shall have a good time defending my theory against people who want to disbelieve the result. I am confident my part of it is right. The world is not so simple as our friends Yang and Lee had hoped. Their main idea still stands and is a permanent step forward of great importance. On Sunday Imme and I sang the Brahms *Requiem*.

TUESDAY, MAY 14, BROOKHAVEN

Yesterday I had a telephone call from [Ed] Creutz in La Jolla saying that he has just been given $10 million by some Texas oil millionaires to establish a big program of research in the fusion energy problem. As a result of this Creutz said he must have me at La Jolla for a week in June to talk over how the program should be set up. I agreed to spend a week in La Jolla before we go to Berkeley. This is a convenient arrangement. I am happy to have the children for a while in La Jolla where we shall stay in a hotel with a swimming pool, and they will enjoy themselves thoroughly. I am also happy about the fusion program which sounds hopeful.

Ed Creutz was the chief scientist at General Atomic, the company that I was working for in 1956. He started a fusion program there that still survives after sixty years. I never worked on fusion myself, but I stayed in touch with the fusion experts at General Atomic and at Princeton.

On Saturday morning I took George to a psychologist to be tested for admission to the township kindergarten next September. This was a serious affair, more like an entrance examination to Winchester than the usual American-style personality test. They kept George busy answering questions for more than an hour, while Esther and I sat outside and read stories to each other. At the end the psychologist had to do a big calculation to find out George's mental age. The reason for all this is that the township decided they will admit children up to six months under the normal age if they show superior ability in such an intelligence test. Esther being born in July got in automatically at 5-5 (five years and five months). George will be 4-6 and so he is a borderline case. It will be a great pity if he is two years behind

Esthi all the way through school. The result of the calculation was that George has a mental age of 5-7. [*His true age at that time was 4-2.*] The psychologist was highly impressed with him and says he has not only a high intelligence but also exceptional ability to listen and concentrate. He said he will give him a strong recommendation to the school board. However, the school board will decide the admissions only in June, and it is not certain he will get in. If he doesn't get in this year, he will have another chance to try for early admission to first grade at the same time next year. George was not at all worried by the whole procedure and also not very interested. When I asked him what questions he had been asked, he replied, "Oh I forgot already."

May 19, 1957

Imme and I just came home from singing the *B Minor Mass*. This is a magnificent work. Unfortunately we were not quite up to singing it decently the first time through. The conductor stupidly refused to cut any of it, so we had no time for a second run-through. Still we improved as we went along, and by the time we reached the Sanctus, we were quite impressive.

The last two days in New York were good, and I feel nostalgic about leaving my job at Columbia. My students gave me a warm sendoff after the last lecture, and I was pleased to see that of the original thirty who had come to the first lecture, about eighty percent stuck the whole course through. That is a good average for such an advanced course. Then on Friday we had a final seminar meeting with Wu and Lee, the Chinese wonders, and a big Chinese lunch to go with it. They spoke generously of the help they had had from my minor contributions. Altogether it has been a good term, and I hope in a few years they may invite me again.

May 30, 1957, Chillicothe, Missouri

Now we are three days and 1,240 miles away from home. The trip goes smoothly. We are not in a hurry and today stopped at five-thirty so we could have a leisurely supper and read a story to the children. The country is green and fresh after heavy rains. This afternoon we crossed the Mississippi, swollen and muddy grey, flowing rapidly to keep pace with the rain. We hear they are having enormous floods to the south in Texas and elsewhere. Up here the floods are comparatively minor. Imme and I drive two hours each at a stretch, and so we do not get too tired. The children are a little bored and have periodical fights and are otherwise charming and amusing.

This evening in the car they started a conversation which kept Imme and me in speechless astonishment. George said, "When I am grown up, I am going to have a station wagon, and I am going to have nobody coming with me. I shall drive across the country, and I shall have just luggage and luggage in the back but no children." And Esthi: "But you know that you have to be careful never to ask anybody to marry you, and then you will not have any children. It is more difficult for me, because some people might ask me to marry them. But anyway I shall never say yes to them." George: "And I shall live in a house all alone and keep the door locked so that children cannot get into it. And when you come to see me, you must knock on the door, and I will look at you through the window and make sure you don't have any children before I let you in." Esthi: "Just think of all the troubles we shall save by not having any children. And so much money we shall save too. Only it is a pity for Daddy because he will never be a grandfather."

I wonder what kind of people these kids will be. At least they start with one advantage.

They are tough.

July 5, 1957

I have been waiting for some definite decision from Verena. So far I heard nothing. I expect a letter one of these days with a final yes or no. I am glad I left the decision to her. In the last resort I am not willing to have on my shoulders the responsibility for separating her from the children. That must be by her own choice if it is to happen.

JULY 19, 1957

Yesterday I got the decisive word from Verena. I think I should explain how this decision came to be made, so that you can see how the responsibility for it is shared. I wrote to her saying that I want her to come back to us, and that if she will come back, I shall give her as in the past my undivided care and love. But I said I will accept her back only on two conditions: (1) that after Kreisel flies back to England on August 22, she will neither see him nor communicate with him anymore; and (2) that she must come back for the reason of her love for me and the children, and not for any reasons of duty or obligations to us. She writes back as follows.

> One thing is clear to me, the conditions you set on my coming back to you are unquestionably the right ones, and I am not able to fulfil them. I think we both know that for me at least to refuse to come back means choosing the harder way. For I doubt very much whether I shall ever overcome the emotional regrets for having left my children and for having left this marriage with you uncompleted. On the other hand I must not come back halfheartedly and I think I cannot do it wholeheartedly. Everything that I can produce so far in the way of rationalization seems to me only a vague approximation to the truth, and therefore I have little to say. I want you to know that you have no reason to worry about me. The crisis I went through this spring is overcome. It honestly pains me to send this letter off, and I hope you do not mind my inability to make a final statement with a heroic bang, though I know that things are growing in an irreversible direction. Love and happiness I wish you. V.

I copy this statement for you verbatim, because it seems to me to sum up in a few words the sensitivity of Verena, the intellectual penetration, and the refusal to compromise her vision of reality with any secondhand sentiments and comfortable simplifications.

SUNDAY, JULY 28, 1957

Such a quiet and cozy Sunday! In the morning Esther expressed a desire to go to church, so the party dress was put on, George was likewise washed and polished up, and the three of us walked down the hill to morning service. Imme said she had enough of Lutheran sermons to last her the rest of her life, so she stayed at home to enjoy a little solitude. I chose an Episcopal church to go to because I like to hear the words with which I am familiar. The church was brand new, built last year in a modern but not extreme style of architecture. The service was exactly as it might have been in any church in Kensington, except that the average age of the congregation was lower (there were a number of children besides mine) and the singing more vigorous. A youngish parson gave a quite intelligent sermon. My children sat through all of it (an hour and a quarter) with perfect decorum and almost complete silence. I found the whole thing restful and pleasant. Though I am not in the slightest degree inclined to return to the religious belief I lost at the age of fourteen, still I feel the children ought to be exposed to it, and the younger the better. It is always impressive to find that so many million people all over America are going to church each Sunday morning with the same King Edward prayer book we learned as children.

After the service we walked merrily up the hill again, and Esther said she wants to go to church every Sunday. But then George said rather apologetically, "I think that for me this church is just a little bit too dull." That was his first and only complaint, after one and a quarter hours of sitting silent on a hard wooden bench.

AUGUST 19, 1957

This morning the kids walked all the way down with me to the office and all the way back alone. This was the second time they had done it. It is only about half a mile, but with many corners and road crossings where cars are all the time passing. Through the summer I have been training them to come with me, each day a little bit further, and now finally they got up to my room and are very proud of themselves. When they walk home alone, I am as anxious as any mother, but I never show them that.

These children are amazing. They can do everything. I am not only a foolishly proud father, it is really true that they are the despair of all the other parents. This morning I drove over the hill to Walnut Creek to watch the children at their swimming class. The class consists of Esther and George and an unfortunate girl called Paula who is the same size as Esthi and is a quite good swimmer. But at the end of the lesson Paula just burst into tears because she couldn't stand the competition.

The real hero of this drama is Katrin. The other day I had a letter from her which quite bowled me over.

Dear Free, I received your letter this morning I am very glad to hear that you and the kids are having some fun even though all thease things are happining. I hope that you and the kids are not too sad about it. Please do not let the kids see that you are unhappy or they shall be more so. I am very sorry that all this happened, but I know that Mommy did what was best. She spent a very long time thinking about it. I hope that she never marries KREISEL (ugg!). I have cried many nights. (I hate all the people who keep pestering me and asking me what is wrong as if I was a baby.) What should I say to them? Is Imme going to stay with you? I hope that she will, she seems to be very nice to the kids. I would like to know what you are working on this summer. Are you working on a problem or are you just hunting? You know that I do not know anything about PHYSICS but I am interested in what my father does because I am very very proud of you. (I know that you are very clever, nice, and that I love you.) Please excuse me for the awkward way that I write things, but I do not know how to write very well especialy when I am nervous. I cannot wait until September. I will be so happy to see you again. Love Katarina.

You can imagine how proud and happy I am to get a letter like this. Now I shall have to do some physics and deserve her high opinions.

OCTOBER 29, 1957

Tonight when I came home from work, it was already dark, and George said, "I have just been outside and I could see Venus and Orion." I did not need to check that his statement was accurate. He has his eyes wide open for all natural phenomena, birds and butter-flies, worms and clouds. Esther is absorbed with her schoolwork and her friends and has less time for looking at the world. On Thursday afternoon I went down to the school to receive Esther's report card and spent fifteen minutes talking with her teacher. The teacher was highly impressed with her and said she is cheerful as well as industrious. Next Thursday I shall have an hour with Katrin's teachers, and that will be a less idyllic picture. But Katrin is working hard and getting good marks, and she is learning to organize her work so it is done with less pain.

Today the whole institute is buzzing with excitement as the rumours go around that our colleagues Yang and Lee are to get the Nobel Prize. I believe this will be announced officially within a few days. It is a wonderful thing for the institute. (Yang is here permanently, and Lee for the year.) It is quite unprecedented for the prize to be given so soon after the work which earned it. But in this case the epoch-making character of the work was understood unusually quickly, so there is no reason why there should be the customary ten-year delay. I feel a little despondent sometimes when I contrast the magnitude of what Frank Yang has done with the barrenness of my own last few years. But I do not feel any worse because of the Nobel Prize. This prize is only the public recognition of a fact which has been clear to us here for some time, that Frank is the brightest young physicist now living. It will cut Yang and Lee off from normal human relations with their friends for some time. They are besieged by newspapermen and hide as much as they can. Luckily there is

Franklin Yang (aged six) and James Lee (aged four) who are good friends to Esther and George and are often running around our house with their cheerful American voices and Chinese faces. The chilling demands of public fame have no effect on them.

One name was conspicuously absent in the announcement of the Nobel Prize. Chien Shiung Wu did not share it. She had done at least an equal share of the work. Like Lise Meitner and Jocelyn Burnell, she made a discovery that would certainly have won her the prize if she had not been a woman. Chien Shiung Wu never complained. She shared the attitude of Jocelyn Burnell, the discoverer of pulsars, the rapidly pulsing radio sources that turned out to be rotating neutron stars. Jocelyn was recently asked by a student whether she was annoyed by this lack of recognition. Jocelyn replied, "Oh no, I feel much better when people ask me why I didn't get the prize, rather than asking me why I did get it."

December 19, 1957

Yesterday I went to Trenton and became an American citizen. Somehow this doesn't seem important. I made up my mind to it long ago.

I was admitted to citizenship together with a large group of immigrants from various countries. The judge who swore us in made a speech, congratulating us for having moved from lands of slavery to the land of freedom. I felt like shouting that in Britain we freed our slaves thirty years before the Americans freed theirs. But I had the good sense to keep quiet when I shook hands with the judge.

14

A SPACESHIP AND A WEDDING

FOR THE ACADEMIC YEAR 1958–59 I took a leave of absence from the Insitute for Advanced Study to work at the General Atomic laboratory in La Jolla on Project Orion. Project Orion was a wild dream, to change the course of human history by flying into space on a grand scale, using our huge stockpile of nuclear bombs for a better purpose than murdering people. We were a bunch of young people who shared the dream. We believed we could actually build a bomb-propelled spaceship with a thousand-ton payload and fly with it to Mars and to the satellites of Jupiter and Saturn. We imagined ourselves cruising around the solar system with our ship and exploring the planets and moons, just as Charles Darwin cruised with the good ship Beagle *around the earth exploring the continents and islands. The dream lasted as a real possibility for about a year. In that year we worked out a detailed design for the ship. We convinced ourselves that it was technically possible to survive the thousands of nuclear explosions that it would take to propel it. We persuaded the U.S. government to give us money to explore the possibilities.*

After the first year, two facts became clear. First, the government would support the project as a research venture but not as a real space mission. Second, the radioactive fallout from the bombs would contaminate the environment to an extent that was rapidly becoming unacceptable. After that year, the project continued to do good technical work, but the dream of a real voyage faded. I went back to my earth-bound job in Princeton and kept in touch with the project only as an occasional visitor.

Forty years later my son George published a book, Project Orion *(2002). He was five years old when the project started and was always interested in its history. His book contains far more information about the project than I ever knew when I was working on it. He examined the official documents that I never saw and interviewed the people who worked on the project and were mostly still alive. He brought the dream briefly back to life. But by the time the book appeared, I had long ago given up any desire to revive Orion. Orion was a dream that failed, a great adventure for all of us who took part in it, and perhaps a model for future dreams that will one day come true.*

At the same time as the big drama of Project Orion, the little drama of my family also reached its climax. The letters in this chapter record both dramas, ending with a marriage in the San Diego courthouse. Between those two dramas there is an interlude at the San Diego Glider Club. The Glider Club owns a landing strip between the General Atomic buildings and the Pacific Ocean. The strip points east-west, perpendicular to the cliffs which run north-south along the shore. The lift that a glider needs to stay airborne comes from the wind blowing up over the cliffs. The lift is strongest over the beach. This means that the glider has to take off upwind and land downwind. The downwind landing is scary when the wind is strong. In 1958 the club-owned glider was a big clunky old two-seater airplane with the engine removed. It came in to land downwind at about sixty miles per hour on a short and bumpy runway. Most of the club members owned smaller gliders that they had built themselves. Those who worked at the Convair factory built their gliders out of parts smuggled out of the factory.

Now, sixty years later, the club is still there, but the old clunky gliders are gone. Instead there are hang-gliders, more beautiful, more convenient, and safer. Next to the landing strip, the elegant modern buildings of the Salk Institute for Biological Studies have given the place an air of respectability that it totally lacked in the old days.

January 1, 1958, Princeton

I promised a report about my physics and about Sputniks. Here it is.

The kind of physics I was doing in 1947–50 when I did my major work was to start with a set of specific equations (supposedly given by God but in fact written down by Dirac, Heisenberg, and Yukawa) and calculate whatever could be calculated. These equations were supposed to predict everything that could be observed. But since 1950 it has turned out that there is no way to get definite results from these equations. Nobody knows whether they have solutions, and there seems no way to find out. So the general belief has grown since 1950 that these equations are mathematically meaningless. The results we got in 1947–50 were good as far as they went, but there seemed no way to go further in that direction.

In 1952 a new point of view was introduced by a young man called Harry Lehmann, then in Göttingen and now spending a year in Princeton. Lehmann's idea was to build up a physical theory without any equations at all. That is, without any "equations of motion" which tell definitely how a physical system behaves as time passes by. Instead, Lehmann allowed only to use three general principles to limit the behavior of the system: (1) the spectrum (meaning that the masses and charges and spins of the particles are given); (2) causality (meaning that no particle can travel faster than light); and (3) unitarity (meaning that the total probability for anything to happen is always equal to one).

Starting this radical program in 1952, Lehmann was only able to prove some weak consequences of his three principles, and it did not look to me then as if any interesting physics would ever come out of it. However, he is stubborn, and he was joined by growing numbers of other people with greater mathematical skill. And now from these three principles we are getting more solid and detailed information about the physical world than we had thought possible. The mathematical difficulties are extreme, and this makes it the right kind of exercise for me. It will be many years before we approach an answer to the central problem, which is to find out how many, and of what kinds, the possible worlds obeying Lehmann's three principles are. I was for many years prejudiced against Lehmann because I thought not enough physics was put into his principles to get anything useful out. It was only this October when Lehmann came to Princeton that I joined in his program. What I have done since is to clean up two mathematical problems that had held him back for a year.

This work is very long-range. It is not particularly affected by the new ideas of Yang and Lee about space symmetry. They have given us new information about the correct "spectrum" to feed into a theory. We are looking at the deep mathematical structure of the theory, which is largely independent of spectrum. My fault for the last three years has been that I worked on second-rate problems to which one could expect a quick answer. Now I have found the subject for a sustained effort.

I have nothing original to say about Sputniks. I feel cheerful about them. It seems to me clear that the Soviet government does not intend to throw bombs at anybody but does intend to dominate the earth by rapid scientific and industrial growth. This will in turn stimulate the Americans to undertake major projects which they would be too parsimonious to do otherwise. There is no question that colonization of the moon and planets will be one of them. I expect eventually to take a hand in this. The prospect seems to me exciting and hopeful.

MARCH 1, 1958, PRINCETON

Katrin was confirmed on Wednesday. She looked fine in a new pale blue dress, white veil, and a pearl necklace which Imme and I bought her as a surprise present the day before. I was glad to be there and especially to go up with Katrin and stand behind her while she knelt in front of the bishop. The singing was good. They take all the hymns at a fast clip and sound as if they enjoy it. Katrin and her friend

Betsy sing regularly in the choir on Sundays. The confirmation service was in the evening, and I took Imme and the little ones along.

The children go on with their lives as gaily as ever. Breakfast table conversation. George: "I know that first there were only ladies in the world, and then afterwards the men came." Esther: "But that is all nonsense. Don't you know that at the beginning there were just two people, Eve and that other guy, what was his name?" Another conversation, showing the difference between the scientific and the practical approach. George: "I can understand how a boat moves along when you push on the oars. You push the water away and so it makes room for the boat to move along." Esther: "But I can make the boat move along even without understanding it."

April 27, 1958, La Jolla

Now I am riding the whirlwind once again. I came here on Wednesday night and had three days of intensive meetings and discussions. Today is Sunday, and I shall spend it alone, thinking over our problems, writing letters, swimming gently in the hotel pool. I have amazing luck in being here [*at the General Atomic laboratory*] just now. Partly it is not luck. The reason the people here are anxious to keep me is that in summer 1956 I was the only one of the famous visiting scientists who wholeheartedly threw myself into the practical problems of building a reactor and came up with some sensible ideas. Incidentally, the reactor we were so hotly arguing about then is now in material existence and will start to operate on May 6. It is a very small reactor, but it has been a good training ground for the people here. They are ready now to build bigger and better ones.

As a result of my reputation for being ready and willing to work on all kinds of problems, I find myself now in a group of people, all of us under forty, planning an enterprise which will inevitably grow into colossal dimensions. The feeling and atmosphere we are now in must be similar to the atomic bomb project in the earliest days, before Los Alamos was thought of, when Oppenheimer and Teller and a handful of other people were feeling their way into the problem and establishing the basic ideas for everything which came later. It is characteristic of this very early time that there is no feeling of pressure or urgency. Everything is informal and relaxed, and we have difficulty in taking the whole situation seriously. In years to come, when huge projects and empires have grown out of this, the early period will have become legendary, and we will not be able to distinguish our memories of this time from the legends which will grow around us. When we shall finally be caught in the glare of newspaper headlines and international political disputes (and this in the end must happen to us as it happened to Oppenheimer), it will seem incredible that the basic plans shall have been made in such a light-hearted spirit as we are living in today.

Since our project never became a huge enterprise like Los Alamos, we never had to face the political storms that engulfed Oppenheimer and Teller. Our project remained active for seven years with a total staff of about fifty people, never attracting much attention, and then quietly disappeared. In the Test Ban Treaty ratification hearings that determined its fate, the project was barely mentioned.

THURSDAY NIGHT, MAY 1, 1958

The word finally came from Washington today: "General Atomic has a contract with the Advanced Research Projects Agency of the Department of Defense to carry out an experimental and theoretical feasibility investigation of a concept for propulsion by nuclear energy of a manned space vehicle capable of interplanetary flight." This statement is not secret anymore, and we are allowed to use it in our efforts to recruit people. However, the Defense Department says we should tell it to people with a request to keep it confidential.

They will be unhappy if it appears in newspapers or is otherwise spread in public. So please keep it to yourselves.

MAY 19, 1958, PRINCETON

I was interested in the reactions to our project at General Atomic. I am something of a fanatic on this subject. You might as well ask Columbus why he wasted his time discovering America when he could have been improving the methods of Spanish sheep farming. I think the parallel is quite a close one. If Columbus had been a patriotic Spaniard, he would have better gone into sheep farming. But he was not serving Spain's purposes. He was using Spain to serve his own. I am unenthusiastic about schemes for irrigating the Sahara. I suppose it will be done someday. But it will be expensive and will take a long time. It is not likely that it could be done fast enough to keep pace with growing population. And it is a problem for politicians and bankers rather than scientists.

We shall know what we go to Mars for only after we get there. The study of whatever forms of life exist on Mars is likely to lead to better understanding of life in general. This may well be of more benefit to humanity than irrigating ten Saharas. But that is only one of many reasons for going. The main purpose is a general enlargement of human horizons.

JUNE 22, 1958, LA JOLLA

I got two good letters from Imme which make me happy. She writes mainly about the children. I had told her by letter the object of our work here. There did not seem to be any need to keep it quiet any longer. Imme replied, "When I drove George to school this morning, I told him about the spaceship. He was very excited, asked immediately which planet you will send it to, and whether there would be a little seat right next to you for him to come along. I had guessed the secret a long time ago and was a hundred percent sure it couldn't be anything else." She also reports, "The kids' report cards are as good as ever. They will be sent off to you, and you will then be able to boast about your offspring on one of the next beer or martini evenings."

June 28, 1958, La Jolla

Los Alamos was beautiful. The air up there at 7,500 feet is so clear and light, now I am down here it feels like treacle. The days in Los Alamos were also exciting. The place is much more alive, and people are much more enthusiastic up there than they were when I went a year ago. I found here a letter from George dictated to Imme. He says at the end, "P.S. I was excited about the spaceship. When we drove home from our big trip we saw a jet plane and a spaceship, and the spaceship was made even before yours. Another love from George."

I never discovered what George had seen that he identified as a spaceship made before ours.

The leader and originator of our project is a young fellow called Ted Taylor, two years younger than I am. He is such a modest and ordinary-looking young man, it is hard to believe when I am chatting with him that this is the Columbus of the new age. But he is it, there is no doubt in my mind about that. He is married to a competent and understanding young wife, and they have four children ranging from ten years to two. I spend a lot of time at their house and enjoy the children. They are not so well-behaved as my children, so you would probably not be too happy in their house. But for me it is fine. On Friday night I went around to their house with a bottle of good cognac, and the three of us, Ted and his wife Caro and I, looked at Jupiter and Saturn through Ted's six-inch telescope. The seeing was good, and we drank to the moons of Jupiter, to our children, and to the success of our enterprise.

Ted was a graduate student during my time at Cornell. I knew him a bit then, but he was violently unhappy with examinations and coursework. He was clearly not intended to be an academic physicist. I did not think much of him in those days. He had been a student at Berkeley but had been thrown out without a degree. He then went to Los Alamos in a very junior position, at a time when all the clever people had left. (This was around 1947.) At Los Alamos he found the people who were left had no ideas and no desire to do anything new. Ted began stirring them up, telling them better ways to do almost everything, until by 1949 a large part of Los Alamos was working on one or another of his ideas. It became embarrassing that he was in a junior position without any degree. So Los Alamos forced him to go to Cornell. He went back to Los Alamos with his degree and continued pouring out ideas. A great part of the small bomb development of the last five years was directly due to him.

In 1956 he came to General Atomic to be head of the theoretical physics division, with complete freedom to do what he liked. He has been producing a lot of good ideas in the reactor business. But he has been living and thinking with bombs for ten years, and he says when he is lying awake at night, he always comes back to bombs. His dream has always been to find some way of putting these tremendous energies to useful work instead of using them for murdering people. Ted got his inspiration last November as a direct result of the Russian Sputniks. He had never been particularly interested in space-travelling. His schemes for peaceful uses of bombs had always been earth-bound and for that reason difficult to put into operation. Then quite suddenly in November he saw that he had the answer to the basic problem of space-flying, how to get enough energy into a sufficiently small weight.

In December Freddy de Hoffmann came to Princeton to tell me of this scheme, and I saw in half an hour that it was the thing all the space-flight projects had been praying for. I have never had any reason to change this opinion. It will work, and it will open the skies to us. The basic idea is absurdly simple. One is amazed that nobody thought of it before. But the only man who could think of it was somebody who had been working and thinking for years with bombs, so that he could know exactly what a bomb of a given size will do. It was not an accident that this man happened to be Ted. The problem is to convince oneself that one can sit on top of a bomb without being fried. If you do not think about it carefully, it looks obvious that you can't do it. Ted's genius led him to question the obvious impossibility. For the last six months Ted has spent his time talking to people in the government and trying to convince them that this idea is not crazy. He has had a hard time. But it seems we have now a lot of influential people on our side. Nothing can stop Ted for long. Ted and I will fly to Los Alamos this evening. We travel like Paul and Barnabas.

JUNE 30, 1958, LA JOLLA

Luckily the work is not regarded as having military importance. I think this is a mistake, but I am happy to leave the generals out of it as long as possible. If the project is successful, they will certainly regret that they did not get into it at the beginning. But for the time being they are not interested. The political problems have been normal and understandable. We asked the government for a few million dollars to get the thing started. The committee which reviews such proposals has at least five hundred proposals a year to look at, most of them crazy or stupid but all of them asking for a few million dollars to get started, all of them submitted by people who get indignant when they are refused. The committee was inclined to say no to us. The thing looks completely crazy at first sight, and they had not time to go into it carefully. We had to wear down their resistance, getting various influential people in the government to believe in us and put in good words for us. The committee has not treated us badly. They gave us a number of meetings to explain what we wanted to do, and in the end they agreed to give us the money. The whole procedure took about six months, but the time was not wasted, as we have been getting on with the work. The preliminary work has been paid for out of General Atomic's pocket.

JULY 13, 1958, LA JOLLA

Such a lovely day it was! My face is as red as a furnace. I was out in the sun and wind for eleven hours, from eight-thirty till seven-thirty. It was a meeting of the Glider Club, and I came out to put the glider together in the morning and stayed to take it apart in the evening. Most of the day I spent doing odd jobs, pulling the towing wire, wheeling the glider around, and so forth. What I like about gliding is that most of the time when you are not flying, there are jobs to do, and it is a friendly group of people. They are mostly simple people who work in the aircraft industry all the week and build themselves gliders at the weekends. The best part of the day was the two flights which I had myself with the instructor. One lasted fifteen minutes and the other ten. I am still very bad, and the instructor is fierce, and I like that. We fly up on a wire which is pulled by a winch, then let go of the wire and sail around in the wind where it rises over the cliffs. This is an excellent place for learning, because the west wind blows steadily up over the cliffs almost all the year round. Today it was a good strong wind, and we could stay up as long as we liked. The seagulls were enjoying themselves in the same wind. It is a beautiful place, with the ocean far below on one side, the yellow cliffs on the other.

I decided I will stay here and bring the family out in September. This means only for one year. We will all go back in the summer of next year. I will have a year away from the institute. It is much better doing it like this than trying to attend to both jobs at the same time. After a year our project will either have collapsed, or it will be so successful that there will not be any great need for me to stay with it.

I talked with Caro Taylor, the wife of our project leader. I was surprised to find her well-informed about English writers of the twentieth century. It seemed odd, because she is an ordinary American girl who married young and has had children to take care of ever since. I found out that her grandmother was a writer called Elizabeth von Arnim, a central figure in the literary society in London fifty years ago. She was for some time mistress to H. G. Wells and afterwards married Bertrand Russell's brother, who is Caro's grandfather. But Caro said that Elizabeth von Arnim's greatest pride was that her grandmother was Bettina von Arnim, who was one of the better-known sweethearts of Goethe.

July 31, 1958, La Jolla

The work goes very well. More people now joined in. We are about twelve altogether. It is very different from a month ago when we were three. It makes me happy to watch the whole thing gradually take shape under our hands, like a figure being chiseled out of a piece of marble. We shall certainly run into difficulties which make us fall into doubts and despair, but so far everything has turned out unexpectedly easy. I am happy I agreed to stay through the first year. One advantage of this is that I am continually flying back and forth across the continent. Last week I had to go east and took advantage of the chance to spend a few days in Princeton. The family was in excellent shape. Katrin seems to love commuting to New York and doing three hours of the most strenuous dancing in the heat of the afternoon. She is outstanding even in comparison with the New York children. Several afternoons I went swimming with the little ones. They were glad to see me, in a relaxed unexcited way.

When I was in Princeton last week, I had a talk with Oppenheimer and got him to agree to my taking a year's leave of absence. He was sympathetic and said he felt a certain nostalgia for the days in 1942 when he was in the early stages of his project. He was emphatic that I should refuse to consider staying away from Princeton longer than a year. It may be difficult when the time comes to step out of this project. I am glad that there will be no choice. Unless I leave the institute permanently, I have to return in September 1959, and I am sure this will be the right thing to do. Princeton was beautiful in the summer heat, especially the nights with hundreds of fireflies flashing in the trees.

August 16, 1958, La Jolla

Last week to Boston where I had a fruitful time discussing our spaceship with the experts. I left behind a lot of enthusiastic people. They have there a variety of "shocktubes," long glass pipes down which they make gas flow extremely fast, and these they can use for doing experiments which will help us along with our plans. Yesterday I had a flight to Pasadena which I visited for the first time. I went to the Jet Propulsion Laboratory, the place where Explorer satellites are made. The reception there was rather cool—the lady at the front office decided Taylor and I were a pair of crackpots and tried to get rid of us. After half an hour of arguing we got inside, and then it all went well.

Today was Saturday, and I went gliding again. It was a lovely day, the wind rather weak but still enough to give some lift on the cliffs. Hot sun and white puffball clouds. I had five flights, and I feel better at it every time. I am not so scared as I was at first. Especially I enjoy the landings, and I am beginning to be able to hit the intended spot instinctively. The greatest achievement today was that I ran the winch all by myself. The winch is the machine which pulls in wire at sixty miles per hour and pulls the glider up into the sky. It is a fearsome machine, you have to jam down the accelerator until it screams at a certain pitch (there is no speedometer) and then let it go gradually up. I was much more scared of the winch than of the glider. But today I gave about ten winch-tows, and the pilots said they were satisfactory. So I am now "checked out on the winch."

August 31, 1958, La Jolla

I have been going through the deeper crisis of my family problem. It has all ended well.

Kreisel is now in Princeton. He is taking Verena away with him. This time there is no coming back. She will go right away to Reno and get the divorce settled. Then they will probably marry. I am not responsible for her anymore. Imme, Esther, and George will join me here in two weeks' time, and we shall start our family life again as it was last year. Verena will drive the car across the country with them. Katrin will choose whether she wants to come here with us or go to a boarding school near San Francisco which has a good ballet school nearby. The people at the New York school where Katrin spent the

summer were enthusiastic about her. They said San Diego is hopeless, and the only good place on the West Coast is this one near San
Francisco. Verena and Kreisel will be living at Stanford where Kreisel now has a job. This will also be close to Katrin's school. So probably she will choose to go there. All this seems to be as satisfactory an
outcome as could have been reached. But Verena had still one more
trick to play. She said she will not deliver the children until I have
agreed in writing I will let them come to her during school holidays.
I do not like to agree to this; there might be circumstances in which it
would be bad for them. I shall probably agree to these terms. I need
these children badly now, and she knows it.

Last night I telephoned and talked to both Imme and Verena.
Imme agreed to come out here for a month or two until the new girl
is settled in. I felt good talking to her again. Then I talked to Verena and called her an unscrupulous old bitch and that felt good too. I
guess that is the end of the story.

SEPTEMBER 17, 1958, LA JOLLA

The family is here, in excellent health, and all have started their
new schools. The drive across country was smooth and rapid. We
are living in a big house with a private swimming pool which is a
great joy. Verena and I are now deep in legal negotiations leading to
divorce. This is a black and horrible business, but I see light at the
end of the tunnel. It would not have been better to do this in New
Jersey. I hope it will all be over in about a week. Margot Kaufner
[*the new au pair replacing Imme*] will be here in two weeks, and Imme
will be gone six weeks later. Imme has been wonderful all through
these lacerating days. Torments and terror vanish as soon as she is
nearby.

SEPTEMBER 26, 1958, LA JOLLA

The nightmare of haggling over our marriage settlement is over.
Yesterday I drove with Verena to Los Angeles, and there we together

signed the papers for our permanent separation. She then got onto the airplane for Reno, and I drove the 150 miles back here alone. I got back at nine o'clock and found the children asleep and Imme quietly waiting for me. The divorce will now take six weeks and is a pure formality, since the difficult questions of custody and finance are decided in advance by the separation agreement. The terms of the agreement are roughly:

1. I have custody of the children during school terms and Verena during the holidays.
2. I pay for them to travel to her three times a year if the distance is less than 1,500 miles, once a year if it is over 1,500 miles within the U.S., and if it is a trip across the ocean, I pay for it only if I am traveling to Europe myself.
3. I am responsible for all support and maintenance of Esther and George; Verena for Katrin.
4. I have undivided possession of the house and the car.
5. I pay Verena $12,000 within two years, plus $1,000 for the lawyers.

Thank God this is all over!

OCTOBER 6, 1958, LA JOLLA

Imme did not say yes to my proposal. She intends to go back to Germany in November and think over at leisure what she will do. I also consider this wise for her. The new girl Margot Kaufner arrived on Wednesday, and she is very good with the children. She will certainly be able to take over from Imme quite effectively. Yesterday the whole family drove into Mexico, and we had a delightful Mexican lunch, Imme and I danced to the music of a jukebox while Margot and the children applauded. Mexico is noticeably more clean and prosperous than it was two years ago.

OCTOBER 7, 1958, LA JOLLA

I am sitting after supper in the big living room of our house while Imme and Margot are chattering away in German. It is good to listen to them, there is such a lot of laughing, though I do not understand most of the jokes. These two girls get along well together, they are both efficient in the house and good with the children. It is a pity I cannot keep both of them!

OCTOBER 12, 1958, LA JOLLA

This weekend I have been alone with Katrin. I enjoy her a lot when the little ones are not in the way. Katrin also enjoys being able to cook and mess around in the kitchen without supervision. She gets my meals, washes my socks, and plays the housewife efficiently. Now she is doing her homework. The family left on Friday morning, so Katrin has the household for three days. I am glad to have some time with her, to talk and get to know her again. She is growing very fast into a young woman.

Yesterday Katrin and I spent the whole day up at the cliffs at the Glider Club. It was a beautiful day. I was in the air five times. Katrin would like to learn to fly, and I will let her begin as soon as I have her mother's permission. But for her the flying is not so important. The reason she will happily spend a whole day up there on the cliffs is Keith. Keith is a fourteen-year-old boy whom I got to know and like during the summer. He is an excellent glider pilot and also a good sport and does more than his share of the routine work. He is too busy with his hobbies to be interested in girls. I had never thought of him in connection with Katrin. However, a week ago when I took her up to the cliffs the first time, she attached herself immediately to him, and he seemed to enjoy her undisguised admiration. Yesterday she spent the whole morning sitting beside him in the cage while he operated the winch, and in the afternoon they went down the cliffs for a long walk together. If this is to be her first serious love, I cannot imagine a healthier or sounder way it could have happened. I never thought her taste was good enough to fall in love with somebody like Keith. As usual, I underestimated her.

Imme and Margot Kaufner and the little ones all went off at six on Friday morning. Katrin was strongly urged to go with them, but she refused. This makes me secretly happy. She said to me, "I like Mummy all right, but I do not respect her." These words were her own, I did not suggest them to her. She knows more about life at thirteen than I did at thirty!

OCTOBER 17, 1958, LA JOLLA

I feel now more optimistic about the chance of Imme saying yes to me. The reason the chances have improved is that I had a reply to my official letter to Dr. Jung in Wertheim. Dr. Jung writes in a very friendly way and says he will not attempt to influence Imme's decision one way or the other. The thing I liked best in his letter was these sentences; [*English translation: "If she decides to become your wife, the difference in age between the partners does not worry me. My own parents lived their lives together in a harmonious marriage, and the difference between their ages was one year greater."*] The problem of the age difference had been for me the greatest worry, and so it was specially good of him to write me those words. Imme also was visibly relieved when she saw the letter.

OCTOBER 21, 1958, LA JOLLA

The good news is here already. Imme said yes. This happened not with any sudden decision but gradually and quietly. Now I am in a state of great happiness and confidence about the future. We have not begun any celebrations because we do not want to tell anybody anything until the divorce is over. Imme insists that not even the children be told before the divorce is settled. I think she is wise; if Verena should know that we decided definitely to get married, it puts Verena into a stronger legal position which she might use to blackmail us further. So please do not say a word to anybody until I give you the go-ahead. The divorce hearing is now set for November 12. Yesterday we had a pleasant hour at a good jeweler's shop looking at rings. I am giving her a beautiful diamond. It makes me happy to spoil Imme a bit. She has never been spoilt before. I promised Imme we shall both come to Europe next summer to visit her family and mine. I wish this news could make you as happy as it makes me.

OCTOBER 26, 1958, LA JOLLA

On Friday night Imme and I had supper at a little restaurant in Mexico. It is a great asset of this place that in an hour one can drive across the border into a completely different world. They gave us excellent food and wine and some Spanish guitar music and took our picture. The picture turned out not so bad as I had expected. We definitely decided on Mexico City for our honeymoon. Dates are still vague, but it will be sometime before Christmas. It is a great help in Mexico that Imme has fluent Spanish. Everything is open to us in a way it never was to me alone.

OCTOBER 31, 1958, LA JOLLA

I came back at one a.m. this morning from a week at Livermore where they make bombs. It is a good sign of my general recovery that I was able to go away for a week and forget my personal drama in the middle of the public drama at Livermore. Livermore was wildly exciting. The days I was there were the last days before the test ban went into effect, and they were throwing together everything they possibly could to give it a try before the guillotine came down. Everybody was desperate and also exhilarated. Edward Teller who is the head of the lab talked to me about his plans. He was in good spirits and pressed me with invitations to come and work for him. But it was good to get home this morning and see Imme and the children again. They had been enjoying themselves and managing things very well without me.

November 6, 1958, La Jolla

George hangs round Imme's neck and asks her please please not to go home to Germany. George is quite worried about her going, and it will be a great relief for him when he learns the truth. Esther is much less attached to Imme and more to her mother. Also she is the only one of the three who genuinely likes Kreisel (and the feeling is reciprocated). Esther is on the surface unworried, but certainly she suffers quite a lot underneath. When she went to Reno for the weekend, she was happy with her mother and found it hard to say goodbye again. When she came home, she said to me, "It would be better not to have a mother at all, as soon as the milk is finished." I think it is best for Esther to see as little as possible of the other family, but I am obliged to send her for Christmas and Easter. She will work her way through these problems somehow.

November 17, 1958, La Jolla

Now for some good news. I heard officially today that the divorce in Reno is decreed.

So now we can tell everybody about our plans. We shall be married here in San Diego on Friday, November 21. I told the children last Wednesday that Imme was going to marry me, and the reactions were characteristic. George: "You know, always when I asked Imme if she was going to stay with us, she always said yes. I believe she has been planning to marry you all the time." Esther: "I am glad there will be two families now. And the children will be able to hop from one to the other, like grasshoppers." Katrin: "Congratulations. I think this is the best thing that could happen for both of you."

November 25, 1958, La Jolla

Just to let you know that we did get married and are still alive. We had a delightful two-day honeymoon in San Francisco, and we shall take off for a real ten-day honeymoon in Mexico City sometime in December. The wedding itself was an unconventional but merry affair. The judge took us between two traffic violation cases. So we all sat, with the flowers and the three children and the two witnesses, in the court while the lawyers were cross-questioning some poor fellow who had been crashed into by a carload of drunk Mexicans. The children were fascinated by the whole proceeding. After a long time the jury came back and pronounced the verdict guilty. The judge then adjourned the case and said, "Well now, let's get these people married." And so we were.

Imme wants a baby very much, so don't be surprised if one is already on the way when you next see her. I was willing to wait a couple of years, but she says she will always feel like Verena's housemaid until she has a real baby of her own. I will be extremely happy if one arrives soon. Next Thursday is Thanksgiving, and we shall have the children here for our celebration.

15

HOMECOMING

In December 1958 Imme and I went to the San Diego airport to fly to Mexico City for our honeymoon. At that time we still had the new au pair, Margot Kaufner, to take care of the children. Imme had a brand-new German passport since her old passport had expired. The Mexican immigration official at the airport examined the passport and refused to let us fly. The expiration date on the passport said clearly that the passport had expired the day before it was issued. We must send the passport back to the German consulate and have a new one issued before we could travel. Four months passed before we could reorganize the honeymoon with Margot no longer available. We finally arrived in Mexico in March.

JANUARY 29, 1959, LA JOLLA

We had a letter from Hans Haefeli. He finally won his appeal to the Santa Rota in Rome. [*This means the pope formally annulled his marriage to Verena. In the eyes of God, as interpreted by the Santa Rota, they were never married.*] He will now marry a Swiss girl whom he has got to know in Boston. Katrin has met her and likes her, but I have not met her. Unfortunately this means he will go back to Switzerland for good. He has a job at a small engineering college in Luzern, and he will settle down there for the rest of his life. Undoubtedly this is the best solution for him, only I am sorry we shall have him no more to visit us in Princeton. He wrote me a very nice letter and said he has only understood, by what happened to me, what an idiot he was for so many years. I am very happy for him that he finally got away from his past too. For Katrin, this does not make much difference. She was not close to Hans anymore. There is no question of her going to live with him in Luzern. It is now entirely up to Verena to make some tolerable arrangement for Katrin.

I am sorry I have to abandon Katrin at this stage. I feel I am punishing her for her mother's sins. But I have really no choice; Verena has legal control of Katrin, and so long as Katrin lives with us, this is bound to be troublesome. Verena's future is also now more settled. She has taken a permanent job at San José State College. This is fifteen miles from Stanford. So she will continue to live where she now is. Kreisel will return to Reading for one more winter, and then come to a permanent position in Stanford in 1960. So it looks as if they will be able to provide a reasonably stable background for Katrin.

Margot Kaufner found herself a job with a family in La Jolla. So she left our household on Sunday. She seems happy at the new place, with five children under ten.

MARCH 26, 1959, MEXICO CITY

So here we are at last! Yesterday I achieved an ambition I have cherished since childhood by standing on the slopes of Popocatépetl [*the gigantic volcano overlooking the valley where Mexico City stands*]. What a magnificent mountain this is! A perfectly symmetrical peak with a crown of white snow. Needless to say, we did not climb it. It is 17,600 feet and takes twenty-four hours and a mule. What we did was just to get near enough to see it in its full glory. This takes some doing. Imme drove us in a beaten-up old Fiat car which we have rented. First she drove fifty miles over a "first-class" road, which means a road which was once tarred but is consistently narrow and full of bumps. Then came twenty miles of "second-class" road, which took us up to the pass about 11,000 feet high. The second-class road is barely endurable if one goes at fifteen miles per hour. It is hard dirt with pockets of soft sand, the surface ridged like a draining board. On the map there are also many third-class and fourth-class roads. But the map says before starting on these one should "seek local advice." We decided the best advice would be to go on a mule. The pass is a magnificent place, a wide expanse of dry plateau with the snow peaks of Popocatépetl on one side and Ixtaccihuatl on the other. We walked up a little hill in the pass and looked through binoculars at the mountains and distant villages far below. Walking there is not an unmitigated joy, because the ground consists of the most penetrating black volcanic ash, which comes through everything and makes people and clothes uniformly grey.

The first days here we spent shopping and sight-seeing in the city. Yesterday was the first with the car. Today we took the car out again and went to see some ruins of pre-Aztec people, date about A.D. 500, at Teotihuacan. These ruins are also immense, and besides colossal pyramids and temples a certain amount of good fresco painting has survived. The pre-Aztecs seem to have been more cheerful people than the Aztecs. All the time here I am reminded of the days when I read Prescott's *Conquest of Mexico* aloud to Mamma. I am glad I have this background to help understand things. But I am still woefully ignorant of the later history.

When my mother was growing old, she had eye problems that made reading difficult. I read many books aloud to her, including Prescott's The Conquest of Mexico *and* The Conquest of Peru *(1843 and 1847). Prescott's books are written in a lively style, with a fast-paced narrative and vivid descriptions of buildings and landscapes. The monuments and mountains that I saw in Mexico were exactly as Prescott described them. I was amazed when I later learned that he had never visited either Mexico or Peru. He had eye problems too, and that made traveling difficult.*

Generally speaking, the city is lovely, full of trees and parks and wide roads with even wider green strips on either side. There are many lovely old and new buildings. And still one is constantly shocked by the extreme poverty in the villages and in many parts of the city, the crowds of people hanging around the hotel district trying to sell knick-knacks, many of them small children. Schools seem to be few and far between. We flew down here on Saturday, shall probably stay ten days and have three more at Acapulco on the ocean before flying home.

On Sunday we went to a bullfight in the big arena. This was not a good bullfight according to the experts, but it was exciting enough for me. The spectacle is impressive, and I should quickly become an aficionado if I lived here. You would not like it here, as the heat at midday becomes oppressive. Even for us it is exhausting from about eleven till three. One has to live like the Mexicans and take a siesta during the hot hours. It is pleasant to find the villages unspoilt, primitive, and picturesque. But for the people who live there, one can only hope they get spoilt as quickly as possible.

APRIL 7, 1959, LA JOLLA

Here we are, safe and sound after our two-week jaunt. The last three days turned out the best. Acapulco is what a seaside resort ought to be. The sea is warm so one can swim as long as one likes. In Acapulco we had no shopping or sight-seeing to do, and we could relax completely. We were in a hotel of the most satisfactory sort, a little bungalow for each family, equipped with icebox, cold drinks, and a jeep for driving around the rugged hills. We found the three children in excellent spirits when we returned. George and Esther stood at the door and threw rice over us as we came in.

APRIL 19, 1959, LA JOLLA

Now comes the sad news. Oliver will not be born in any inn or stable. Imme went to see her doctor, and he said Oliver is due approximately September 6. He said it is crazy to think of coming back from Europe in August, and even the trip out in July would be a serious risk in case of a bad-weather flight. So we decided to be sensible and have the baby here. We will drive home in time for the start of the institute term on October 1. It was Mrs. Peierls who said, "It is no use trying to have a baby at a convenient time. When they come, it is always inconvenient."

Each of our daughters in turn was called Oliver, until he turned out to be a she.

Two breakfast conversations, one from yesterday and one from the day before.

Freeman, holding forth as usual like a professor: "Scientists are still trying to find out how pigeons can find their way home. We don't know whether they can find their way by the stars." George: "But I know that camels find their way by the stars." Freeman: "How do you know that?" George: "Because when the three wise men were going to Bethlehem, the camels saw a star in the sky, and so they knew where to go."

Esther is looking through an anthology of children's poems: "That's funny, I wrote a poem called 'Daisy Song,' and now I see another author has written one too." Freeman: "Who is the other author?" Esther: "John Keats."

MAY 19, 1959, LA JOLLA

We have a nightingale who sings each night outside our bedroom from about midnight for three or four hours. He stops just before sunrise when the chorus of other birds begins. He sings so loud it sometimes wakes us up, but it is no hardship to lie awake and listen to him. He sings alone, and it is impossible not to imagine a wealth of feelings behind this outpouring of song.

A Russian astronomer, Iosip Shklovsky, has suggested halfjokingly that the two moons of Mars are hollow. There is some evidence for this. The inner one has changed its revolution period faster than it should if it were solid. If they are hollow, then they are artificial, and the question is "Whose?" Shklovsky is a good astronomer and has done many sensible things before this. Now I am more than ever impatient to go there and have a look around.

SUNDAY, MAY 24, 1959, LA JOLLA

We just came back from a Sunday excursion to the Palomar telescope. This is only a two-hour drive from here, and it is through lovely country all the way. The observatory is 5,550 feet up, and the cold mountain air was most welcome to us. Also there is a little museum with thirty of the most striking pictures taken by the big telescope on view, properly illuminated with light coming through from the back. One never sees them so well when they are printed on paper. The big telescope, which started operations in 1949, is still unique in the world and still producing a large proportion of the important astronomy. It is surprising that nobody is thinking of building a bigger one. Even a second one of the same size would be enormously useful.

The astronomers are seriously interested in putting telescopes into space. Martin Schwarzschild in Princeton is a good friend of mine, and last year he flew a twelve-inch telescope several times in a balloon to eighty thousand feet, to take pictures of the sun. He had the telescope point itself automatically at the sun and take the pictures automatically. It all worked well, and he had some pictures sharper than any that had been seen before. Schwarzschild said the problems of operating a telescope in a rocket satellite are in some ways easier. He is already getting ready to send up a small telescope into orbit. I will be surprised if this is not done successfully within two years. I am interested in putting a bigger telescope higher up, for example on the moon. One of the attractions of our spaceship is that it could carry a big telescope up there with enough people and stuff to use it effectively.

After Einstein died in 1955, Niels Bohr was the greatest living physicist. He presided over European physics from his home in Copenhagen. Unlike Oppenheimer and most of the other senior physicists, Bohr believed strongly in the peace-

ful uses of nuclear energy. Freddy de Hoffmann invited Bohr to come to San Diego to give his blessing to the formal dedication ceremony of the General Atomic laboratories, and to our amazement Bohr said yes. He stayed at General Atomic for a week and was seriously interested in the details of our work. He was particularly interested in the TRIGA reactor and the Orion spaceship. As part of the dedication ceremonies, he pushed the switch to demonstrate the safety of our prototype TRIGA. The switch caused the control rods to be pulled explosively out of the reactor, so that the reactor was supercritical with a neutron doubling time of about two milliseconds. For a few milliseconds it was blowing up like a bomb. This was the worst possible situation that could happen in a reactor accident. The reactor power shot up in a few milliseconds to fifteen hundred megawatts, enough power to melt down the reactor in a few more milliseconds. But there was no meltdown. Instead, the power came down as fast as it went up, and the reactor settled down quietly to a power of one megawatt. The audience heard a loud bang when Bohr pushed the switch, and then silence. They were invited to walk by the reactor in its swimming pool and see the blue glow in the water around it. The automatic shutdown of the power after an accident was guaranteed by the laws of physics and did not require any human intervention. Bohr enjoyed the reactor like a boy playing with a new toy. He said he was sorry he was not able to include the launch of a real Orion spaceship in the dedication ceremony.

Yesterday we had a beautiful supper-picnic on the beach. A lot of our friends were there and also Niels Bohr and his wife. Old Bohr was here for the official celebrations when the laboratory was dedicated last week. Bohr stayed another week to talk to people and enjoy the scenery. For about half an hour Bohr talked to me alone as we walked up and down the beach. It was a tantalizing experience, as his voice is almost too low to hear, and each time a wave broke, his wisdom was irreparably lost. I learnt a lot about his struggles during the war to convince Roosevelt and Churchill that the atomic bomb was not something they could keep in their pockets. He believed then that the Russians would soon be making their own bombs and that the only chance to avoid a catastrophe was to bargain with Russia immediately for an abandonment of secrecy on both sides. Of course, he failed to convince Roosevelt and Churchill. But he said he had General [Jan] Smuts and Lord Halifax on his side.

I was glad that Bohr was enthusiastic about our spaceship. He thinks of it as something with which one may once again try to make a reasonable bargain with Russia. Of course, the difficulties now are in some ways greater than they were in 1944. But at least the secrecy problem is not so obsessive as it was then. The politicians have learned that secrets do not stay secret forever, and one can talk much more freely than in 1944. The problem is that we do not have much to offer Russia in return for opening up their country to us. I do not have much hope that we can solve the problems of war and peace this way. But I am glad if we try. Bohr encouraged me a lot.

September 7, 1959, La Jolla

I got the children finally into bed. The house looks like a battle-field, but I will not try to tidy it up until Imme comes home. I take a few quiet minutes to write of the events of the day. It began at one a.m. with Imme gently waking me up. She said Oliver had begun hurting her around midnight, and it was now starting to feel seri-ous. I timed the pains and found they were regularly at six-minute intervals. At one-thirty the interval changed to three minutes, so we telephoned the doctor and drove over to the hospital. We got to the hospital about two and found a great deal of commotion since three ladies had arrived in rapid succession. Because Imme had a first baby, they thought she would be slow and left her until last. However, about four they suddenly found out she was reaching the crisis, her doctor was summoned from his bed, and he arrived just in time to deliver a five-pound girl at 4:47. [*At that moment Oliver became Doro-thy.*] The birth was quite simple and the baby started to scream at once. I was with Imme for a few minutes just before the birth, when the pains were severe. I came back an hour later and found her smil-ing and relaxed, everything over and already half forgotten. At six I drove home and had a fine time telling the children all about it. They were happy but not specially excited. They had been expecting this to happen each night since they arrived. Esther said she wants to marry young so that she can start having her own babies soon. This remark is a welcome change from her expressed intention to have no children at all, which she often used to assert when her mother was grumbling about the burdens of home and family.

George said when I put him to bed, "I am so excited about going to Princeton that I can hardly sleep. And I am excited most of all about the deep growly snowstorms."

December 18, 1959, Princeton

Last week I spent writing a political article about the problems of nuclear weapons. I hope to publish it in a quarterly magazine called *Foreign Affairs* which reaches a select and intelligent audience. The editor of *Foreign Affairs* is a young man who lives in Princeton. He brought off a tremendous success this year by getting a fifteen-page article by Khrushchev. He wrote to Khrushchev shortly before he came to America asking for an article, and back it came within a few weeks. The experts said it was undoubtedly written by Khrushchev because of details of style and phrasing. My article has to go to Livermore first to be scrutinized for secret information. When I get it back, I will send you a copy.

December 31, 1959, Princeton

It is good to hear that you are enjoying Venus in the morning skies. Did you see in your newspapers that they have measured the surface temperature of Venus by looking at radio waves coming up through the clouds and they find an average temperature of 575 Fahrenheit? This was done at two different observatories independently, and the result is presumably correct. We had guessed it would be 100 or at most 150. This is an interesting puzzle, to understand why it is so hot. It is sad for the people who would like to go and see the place.

Even at the poles one probably has an atmosphere consisting largely of superheated steam. This puts Venus permanently out of the running for colonization.

Later, when I looked more carefully at the problem, I concluded that colonization of Venus would not be impossible. If we put a big sun-shade in orbit around Venus, the atmosphere and the surface of the planet would cool down to a comfortable temperature in about five hundred years, and colonization could then begin. After the cooling, the main obstacle to colonization would be lack of water. That could be alleviated by importing water from icy asteroids or comets.

APRIL 7, 1960, PRINCETON

By good luck my article in *Foreign Affairs* appeared this week when everybody was interested in the subject. I had some busy days, with reporters calling from Washington and photographers coming to the institute to take my picture. I am pleased with the way this whole affair went. I have made a number of enemies by saying things people wish to be unsaid. But I now have a much easier feeling in my own mind. Whatever the government decides to do, I have done what I could to push them in a sensible direction. I do not have to feel responsible if my advice is not taken. I can now keep quiet and go back to working at physics with a good conscience.

The Foreign Affairs *article was a political diatribe, arguing against an international agreement to stop the testing of nuclear weapons. I was emotionally opposed to a test ban treaty because it would forbid the nuclear tests that Orion would need before it could fly. When the Test Ban Treaty was signed in 1963, I had changed my mind and gave testimony supporting the treaty at the Senate ratification hearings. The Senate ratified the treaty with my blessing. By that time, I had given up hope that Orion could fly, and I was happy to join the majority who supported the test ban as a way to slow down the nuclear arms race. Orion was already dead, and the treaty was only one more nail in its coffin.*

One surprise still remained for me. In December I showed my article to Oppenheimer before it was accepted by *Foreign Affairs*. I did this so that he could object to anything he found objectionable. He surprised me then by agreeing to everything and saying it was all right with him. Now after it appeared in print and some of his friends complained to him about it, he violently changed his mind. He gave me half an hour of tremendous scolding, saying that it was wrong to make any public statements about technical matters unless the statements were of such a kind that all experts would support them unanimously. This happened at a party in Oppenheimer's house, while Imme stood by and listened in amazement. I shall certainly not follow Oppenheimer's advice, nor has he usually followed it himself.

This is to me interesting, mainly in showing how unstable a character Oppenheimer is. This has always been at the root of his troubles. I am not worried for myself; it is now the third time that I had a tongue-lashing from Oppenheimer, and each time we remain friends. The first time was when I first came to the institute in 1948 and he objected to my physics. The second time was five years ago when he accused me of betraying some faculty secrets. On that occasion I really was at fault. Anyhow, with all his inconsistencies, I like him.

Yesterday I met Frank Drake, who is in charge of the project in West Virginia listening for radio signals from inhabitants of nearby planetary systems. So far they have tried two stars. Drake is a first-rate man, absolutely sound and full of common sense. He was here to talk to the astronomers about other things he has been doing, exploring the galactic center with radio. He said the big problem with the listening project is to be on guard against teenage pranksters with radiotransmitters (of which there are thousands in this country alone) who might enjoy fooling Drake with some plausible signals. Drake had a signal which seemed to be genuine for a whole month before he finally proved it came from Earth. Fortunately this did not get into the newspapers. My scheme of looking for unusual infrared emissions would alleviate this problem.

My proposal was to search for alien societies who were not interested in communication and could not be detected by radio. Any advanced society in the sky, whether or not it wished to communicate, could not avoid radiating waste heat into space. We could detect any large emission of waste heat with a telescope sensitive to infrared radiation. Many years later, telescopes in orbit searched for artificial sources of infrared radiation, but found no evidence for high-powered alien societies.

MAY 24, 1960, PRINCETON

I have occupied myself for the last few weeks with minor things. One of these was a little paper on the detection of intelligent beings in the universe [1960]. This was rather a crazy idea, but I think it is fundamentally right. I have still a strong belief there is a good chance we shall find evidence of intelligence in the universe if we look for it hard enough. This is not necessarily a remote prospect—it could happen within the next few years. I have also the naïve feeling that if we do find something of this kind it will have a profound effect on human behavior in general. It has always been true historically that the way to unite factions has been to find a common enemy. So the discovery of some rival civilization out there would have more effect in uniting humanity than any number of summit meetings. That is my belief. So I am in my own way fighting for peace too.

In August Imme and I traveled with our eleven-month-old Dorothy to England and Germany to meet her grandparents.

AUGUST 31, 1960, PRINCETON

Our trip home was comfortable, but we were three hours late in New York. In New York we missed all the convenient trains, and so we went to the bus station to wait for a bus to Princeton. At the bus station one could buy a quart of ice-cold milk, and so I took Dorothy's bottle to the Men's washroom to wash it out before refilling it. I felt rather foolish in the men's room with this baby bottle and two huge soldiers washing their hands on either side of me. However, one of the soldiers after looking at me said, "Say, that's not the right way to clean out a nipple. I'll show you how to do that. It's something I'm a real expert at." So he showed me the expert way to do it, and I felt, well, this is America, and I am glad to be home.

16

WORKING FOR PEACE

In 1960 I became involved in problems of war and peace on two levels, inside and outside the government. On the inside, I was invited to join JASON, a group of scientists doing technical studies for the U.S. government. Since I had already been cleared for nuclear information at General Atomic and Los Alamos, I had no difficulty getting cleared for work with JASON. The main concerns of JASON at that time were the security of our strategic forces against surprise attack and the feasibility of defense against ballistic missiles. Jason still exists and has been giving advice to the government about technical problems for fifty-five years. We have not had much influence on the big questions of public policy, which are political rather than technical. We were not asked whether the invasion of Iraq in 2003 made sense. Our most useful contribution has been to kill wasteful or fraudulent projects that tend to grow wherever funds are available. The rules of secrecy make it easy for artful dodgers to waste money. Secrecy hides failure much more often than it hides success. Members of JASON have the clearances that allow us to poke around in dark corners.

On the outside, I became active in the Federation of American Scientists, a political organization of scientists pushing for international agreements to slow down the nuclear arms race. To my surprise, having only recently become an American citizen, I was elected in 1960 to the council of the federation. The image of America as a melting pot, in which aliens are absorbed and respected, has remained for me a reality, I quickly became a friend of Daniel Singer, the general counsel of the federation, a lawyer who lived in Washington and organized the activities of the federation under the guidance of the council.

My entry into American public life coincided with the start of the Kennedy presidency, a time of rapid change and youthful rebellion. In retrospect, the 1950s were a time of stability, discipline, and old-fashioned loyalty. The 1960s were a time of turbulence, violence, and new ways of thinking. In the 1960s, American society was torn apart by the war in Vietnam. Kennedy started the American involvement in the war. After Kennedy died, President Johnson escalated the war in spite of a rising tide of opposition. After the 1960s, President Nixon had the wisdom to accept defeat. This is the picture of the 1960s that we mostly see today. My letters show a more nuanced picture. The tragedy in Vietnam and the violent conflicts in America did not dominate our lives. The quiet efforts to create a more peaceful world and a more just society continued through the 1960s. These efforts are the main subject of the letters.

NOVEMBER 1960, PRINCETON

On Tuesday I voted for Kennedy. That night we went to a neighbour's house to watch the election results on television. Everybody among our friends was on the Kennedy side. Hardly a single word in favor of Nixon had been heard in our house. So I was surprised when I asked Esther and George, and they said they would both vote for Nixon. When I asked them why, Esther said, "Because he has been vice president and so he has had more practice." George said, "Kennedy is against science fiction movies." We are happy this election went to Kennedy. There seems to be a chance he will put an end to the penny-pinching which has paralysed a lot of things here in the last few years.

JANUARY 4, 1961

Imme and I became good friends with Helen Dukas, Einstein's longtime secretary and executor. As a result I have been reading the volume *Einstein on Peace*, which was just published. I find it impressive. It is a collection of his political writings since 1914. What is remarkable is that he wrote so much and so well about political matters, even at a time (1914–18) when he was prolifically creative in physics. He was as outspokenly opposed to Kaiser Wilhelm in 1914 as he was to President Eisenhower in 1954.

Helen Dukas lived with Einstein's stepdaughter Margot in the Einstein house in Princeton. After Einstein's death she continued to work every day in the Einstein archive on the top floor of the institute. She loved children, having grown up in a big family and never having children of her own. Imme invited her to our home, and she quickly became a substitute grandmother for our children. She was their favorite baby-sitter and remained a close friend until her death in 1982.

JANUARY 20, 1961

I made an expedition, technically known as a junket, a pleasure trip disguised under a serious purpose. The JASON group, the group concerned with scientific military problems, spent three days at Key West, to see submarines and submarine chasers and to learn about their practical difficulties. The navy staged a day of submarine hunting for our benefit. I was on the destroyer *Huntington*. We watched the sailors busy with their radars and navigation charts. It betrays no military secrets to say that the submarines did better than we did.

FEBRUARY 12, 1961

On Saturday, February 4, we had our big blizzard. I had two meetings in New York that day. In the morning I was chairman of a session of the American Physical Society. In the afternoon we had a council meeting of the Federation of American Scientists, the political organization which tries to push the government into doing reasonable things where nuclear weapons are concerned. At seven a.m. I stepped out into deep untrodden snowdrifts and trudged the one and a half miles to the station. Luckily the trains were running and I arrived in New York in time for the nine-thirty meeting. In the afternoon we had the federation meeting which was much more interesting.

They are a pleasant group of people, a bit leftish in politics and intensely well-meaning. The one subject on which I disagree with them is the test ban, which I oppose and they have made a prime objective of their campaign. The meeting started predictably with a discussion of the test ban. Many of them spoke suggesting ways of getting the public more enthusiastic about the test ban. Then it was decided to delegate to the executive committee the job of making some more vigorous test ban propaganda. At this point I decided to speak up. I said they could do whatever they liked about the test ban but that I considered they were wasting a disproportionate amount of effort on it. I said that to me the general problems of disarmament and the use of the existing weapons seemed hundreds of times more important than any test ban. So they did move on to talk about disarmament. They talked a long time and agreed to pass a long resolution pointing out the desirability of general disarmament. It was a fine resolution except that it didn't say anything specific. At this point I again made a speech saying that I was unsatisfied with vague generalities, and the federation ought to be discussing some real proposal to change drastically the existing international dangers. They replied, "What do you have in mind for us to do?" And I said, on the spur of the moment, not having anything prepared, "Let us see first of all whether this council can agree or disagree with the following statement:"

We urge the government to decide and publicly declare as its permanent policy that the U.S. shall not use nuclear weapons of any kind under any circumstances except in response to the use of nuclear weapons by others. We urge that the military plans and deployments of the U.S. and its allies be brought as rapidly as possible into a condition consistent with the over-all policy of not using nuclear weapons first.

I was rather taken aback by the response to this. It was over-whelming. I had myself been feeling for some time that our greatest danger comes not from having nuclear weapons but from being com-mitted to using them in stupid and disastrous ways. To most of the council, this seemed to be quite a new idea. Not one of them spoke seriously against my proposal. In the end it was voted on and carried unanimously. So at one stroke my reputation as a reactionary war-monger was destroyed, and I became a shining champion of peace.

It remains to be seen what impression this action of the federation will make upon the public. It could conceivably be important. The federation is not as influential as it would wish to be. But we have good connections with people in high places. Our meeting went on till midnight. All the time the snow was swirling past the windows. I took the 12:05 train to Princeton Junction. In return for helping to dig somebody's car out of a snowdrift, I was driven back to Princeton. I arrived home weatherbeaten but exhilarated at three a.m.

I wrote a longer statement for the federation, explaining the arguments for and against the No First Use policy. Our statement, in both the short and the long ver-sions, was distributed to the newspapers and news agencies. On the following day the newspapers were still filled with stories and pictures of the great blizzard. There was nowhere any mention of No First Use.

MARCH 1, 1961

My next-door neighbor George Kennan is now preparing his departure. He will go in April to be ambassador in Belgrade. In the meantime he has been to see Kennedy and talk over with him the gen-eral problems of dealing with Russia. Kennan is full of enthusiasm. He is happy that he is again back at the job for which he was trained. Also he found Kennedy personally electrifying. It seems Kennedy has this effect on many of the people who meet him. We were lucky to have voted him in.

I never had any direct contact with Kennedy. Besides being a friend of George Kennan, I also got to know Arthur Schlesinger, who came as a visiting member to the institute in 1966. He had been personal assistant to Kennedy from 1961 to 1963 and wrote many of Kennedy's speeches. He published a book, A Thousand Days: John F. Kennedy in the White House, *with vivid accounts of Kennedy's electrifying effect on people who came to see him.*

Kennan had been declared persona non grata by the Soviet government when he was ambassador in Moscow in 1952. He then came to the institute to start his career as a historian. He took a leave of absence from the institute to be ambassador in Belgrade, hoping to improve relations between Yugoslavia and America and incidentally strengthen the independence of Yugoslavia. For various reasons these hopes failed, and in 1963 Kennan came back to Princeton and the institute permanently, saying he was glad his days as a diplomat were over.

APRIL 29, 1961

A hectic week in Washington. Four days of physics meetings, three days of JASON meetings, and two night meetings of the federation. The federation meetings were the most interesting. There were people there who are working in the president's staff on disarmament proposals. They said my memorandum on No First Use has reached the desks of important people and is being used as a working paper. The time I spent in writing and arguing for it was not wasted. The council did not take any important new action. We are waiting to see what Kennedy will do when the big disarmament negotiations begin again in September. His staff is preparing positions for these negotiations. It seems there is a chance something solid may come out of these efforts.

People in Washington are all excited about the Cuba fiasco. It seems to me not so tragic. I never believed the invaders could succeed, and I am glad the affair was over quickly. I think the effect on Castro has been good, he is now more sure of himself and can afford to be somewhat relaxed.

The Cuba fiasco of 1961 was the Bay of Pigs invasion of Cuba, not the missile crisis which happened a year later. The Bay of Pigs invasion by a small army of Cuban exiles was a dismal failure. Kennedy gave the invaders moral support but wisely refused to reinforce the invasion with American troops when it got into trouble. I misjudged the effect of the invasion on Castro. Instead of becoming more relaxed, he became more bitterly hostile to the United States and invited Nikita Khrushchev to install Soviet nuclear missiles in Cuba. The resulting missile crisis ended peacefully, with both Khrushchev and Kennedy acting with remarkable restraint. The missile crisis is not mentioned in the letters. Everyone was scared to death when it began and hugely relieved when it ended. The narrow escape from disaster helped both sides to accept the Test Ban Treaty that was negotiated in 1963.

May 25, 1961

You ask what I have been calculating so industriously for the last three months. I will try to explain what it is about. The idea is to work out a new kind of statistical theory which will apply to the dynamics of heavy nuclei.

I neglected to mention in this letter that the new kind of statistical theory was invented by Eugene Wigner at Princeton University ten years earlier. What I did was to take Wigner's physical idea and work out the mathematical consequences in detail. As a result of my work, the theory became more precise and also more general. It could be directly compared with experimental data and extended to other physical systems besides heavy nuclei.

You may know that the statistical method has been successful as applied to ordinary gases. Since it is impossible to predict the motions of the molecules in detail, one goes to the opposite extreme and assumes complete ignorance of the state of the molecules. This assumption of maximum ignorance leads to definite laws of behavior for gases, which are in agreement with observation. The theory was worked out by James Clerk Maxwell and Willard Gibbs about a hundred years ago. Now my idea is to carry this method to a deeper level. In the nuclear problem one is ignorant not only of the state of the particles in the nucleus but also of the laws of force under which they move. So I developed a new statistical mechanics in which all possible laws of force are regarded as equally probable. The main problem was to make this notion mathematically precise. After that it was a job of calculating various quantities that could be compared with the properties of actual nuclei. The calculations are tough, but this is just the kind of thing I know how to do. The agreement between theory and observation is close but not perfect. There is a lot still to be done. The observations which I explain are done at the big research reactors at Brookhaven, at Chalk River in Canada, and at Harwell. There is a lot of beautiful data, and the experimenters are happy that someone is at last taking some interest in it.

The initial step in this work was taken by a young Indian student called Mehta working in France. We have invited him to Princeton, and he will be here in December. It was after reading his papers that I started thinking and was able to carry his methods much further.

Madan Lal Mehta came to the institute several times and became my chief collaborator for many years. We published papers together, and he wrote the classic book, Random Matrices *(1967), summarizing my work as well as his own. He was a great traveler. His last adventure, an overland trip from Beijing through China and Tibet to his home in India, was undertaken after his retirement from his job in France. The trip included a week in jail in Tibet and a walk over the Himalayas. He wrote a graphic account of his journey, which remains unpublished because it might endanger some of the people who helped him.*

FEBRUARY 12, 1962

Yesterday I got a sudden phone call asking me to stand for election as chairman of the Federation of American Scientists. This is the political organization which I joined two years ago. And now they want me to be chairman. I am astonished, so many of the others have been working hard in the organization for fifteen years, and I have never done anything substantial in the two years I have been with them. It is an opportunity to make my opinions heard and to start things moving in new directions. I have no doubt I ought to accept the chance. It will mean a lot more responsibility than I ever took on before. More trips to Washington, more long-distance telephone calls, all the things I dislike most. But this time it is in a good cause.

APRIL 10, 1962

Today a telegram came from Washington: "Congratulations on your election as chairman of the Federation of American Scientists." Now I must start thinking seriously about what to do with my year of prominence.

APRIL 18, 1962

I am flying home from Notre Dame University. The Notre Dame mathematicians organized a small conference to discuss problems on the periphery between mathematics and physics. It happens that there are several Polish mathematicians at Notre Dame, and they invited all the Poles within sight, so there was a grand reunion of Poles who had emigrated at various times with others who are visiting from Poland. The Polish mathematical school has a special flavour, which has survived remarkably well the vicissitudes of the last thirty years. There was a great deal of talk about Infinite Games, which are the latest fashion in Poland, and which clarify in a remarkable way some of the deep questions in the foundations of mathematics. I found myself happily arguing about these questions, which I had last thought about seriously when I was a student in Cambridge twenty years ago.

The city of Lwow was the most vibrant center of the Jewish community in prewar Poland, including writers and rabbis as well as mathematicians. Those who did not emigrate in time were almost all massacred by Soviet, German, and Ukrainian occupiers. The city is now called Lviv and belongs to Ukraine.

My friend Stanislaus Ulam came to the meeting from Los Alamos. He had been one of the most brilliant young men in the golden age of the Polish School (1930–39). During the war he went to Los Alamos to work on bombs and he has stayed there ever since. He is one of those people who pour out wild and original ideas and leave it to others to fill in the details. One of his ideas was the starting point which led Teller to invent the hydrogen bomb. Another of his ideas was the bomb-propelled spaceship which later grew into our project Orion in La Jolla.

Ulam told me that it happens to him from time to time, when he is out of doors in the evening and the light is of a particular quality, that he suddenly begins thinking about problems in the foundations of mathematics. This happens because the evening light used to be just like that in Lwow when he wandered through the streets as a boy, before bombs and spaceships came to distract his attention into more mundane directions. For me the foundations of mathematics have the same sort of nostalgic associations. So we lived at Notre Dame for these four days in a continuous conversation, half in Polish and half in English, drinking immense quantities of black coffee. Stimulated by the atmosphere, I gave one of the best talks I have ever given. Several people came to me afterwards and said that Ulam and I were the high points of the meeting. Notre Dame University is a Catholic foundation, and most of the senior administrative jobs are held by priests. One of my mathematician friends [*Paul Erdos*] said, "Notre Dame is a very good place, only there are too many plus signs."

April 26, 1962

It has been a remarkable experience to be the chairman of the Federation of American Scientists. We are such a small group of people (two thousand members all together) and we are mostly concerned with FAS only in our spare moments. So it is astonishing to discover how seriously the people high in the government take our opinions. My office seems to be a key to open all doors. In three days I have been in turn to talk with (1) the second-in-command of the Space Agency, (2) the second-in-command of the Disarmament Agency, and (3) Walter Reuther, the boss of the United Automobile Workers. All these interviews were arranged by our man in Washington, Daniel Singer, who is the organizer of our activities. Chairmen come and go, but Dan Singer remains. The reason why we have such an influence is mainly that we have used our influence wisely in the past. For example, last year the FAS put effective pressure on Congress by convincing a number of congressmen to establish the Disarmament Agency. Naturally the people who are now running the Disarmament Agency are grateful to us.

I saw the Space Agency people mainly to appeal to them to put more money and effort into university research and student fellowships. In particular, I want them to build several big telescopes of the Palomar type. One such telescope costs only about as much as two of their big rockets. They replied that they do as much of this kind of thing as they dare, but Congress does not understand it and cuts their supply of money if they try to use it for general educational purposes. This is partly true, but they are much too timid in dealing with Congress. We agreed that they shall push their university program harder while we meanwhile try to educate the congressmen.

With the Disarmament Agency man (Frank Long) I talked mainly in my private capacity, in order to arrange to work in his organization during the summer in the most effective way. It looks as if I can spend two months in his office in Washington, where I am also in a good position to carry on FAS activities.

The most impressive by far of these important gentlemen is Walter Reuther. He is a successful union leader (with 1.25 million men in his union) and at the same time an intellectual and a social philosopher with all kinds of ideas for the reform of society. Roughly he is a combination of Ernest Bevin with Aneurin Bevan. [*Bevin was a British union leader, Bevan a left-wing politician.*] He had a big part in getting Kennedy elected president (the UAW put all its muscle behind Kennedy's campaign), and he now is able to talk to Kennedy with great freedom. He also spent two years in his youth building a car factory in Russia. (The Russians bought the tools from the Ford company and Reuther went over to teach the Russians how to operate them.) He has strong views about Russia and gave Khruschchev a bad time when K. was invited to supper with the union leaders in 1960.

Reuther is now deeply concerned about disarmament, understands that disarmament is essential, and is trying to get the government to make plans so that disarmament can be done without throwing half his men out of work. Reuther is convinced that this can be done if only the government is not afraid to face up to the size of the problem. We agreed on certain measures of collaboration so that his union can act as a channel for some of our information. Altogether very encouraging.

July 30, 1962, Washington

Yesterday was cool, and I decided to make the most of it. I walked across the Potomac to the Arlington Cemetery and walked around there for some hours in the sunshine. The main thing to see is the old house of General Lee, a lovely place built in the style of Mount Vernon, high on a hill overlooking the city of Washington. The house is kept as a museum, in a sentimental nineteenth-century style, with the children's toys laid out on the floor of the playroom. I find it more appealing than the palatial eighteenth-century style of Mount Vernon. The cemetery is tastefully laid out on the brow of the hill, with many trees to give shade and privacy. The most pleasant thing about the graves is the occasional Samuel Birnbaum or Ezekiel Rosenblum with his Star of David breaking the line of crosses.

In the evening I went to a bookshop to find a collected edition of Wordsworth. I wanted a motto for a study which I am writing on the changes which are likely to occur in the future in the fields of strategy and disarmament policy. I found what I was looking for:

Drop, like the tower sublime
Of yesterday, which royally did wear

His crown of weeds, but could not even sustain
Some casual shout that broke the silent air,
Or the unimaginable touch of Time.

—WORDSWORTH, *Sonnet on Mutability*

This study which I am writing is bound to be only a sketch, since I am here such a short time. I might later expand it into something more substantial, and then I would write it in such a way that it might be read by the general public. A friend of mine called Amitai Etzioni has recently written a book of this kind, *The Hard Way to Peace* [1962]. He is a young professor at Columbia. I find his book the best thing that has yet been written on this subject. Though he makes technical mistakes, he has a largeness of view, and a willingness to face up to all the difficulties, which are very rare. I read in the newspaper that Etzioni had been in Geneva and gave a lecture to the American and Russian delegations at the disarmament negotiations. I hope they listened to him. The first time I ever heard of Etzioni was when a visiting disarmament expert from Moscow told me I ought to read his book. It seems to have made more impression in Russia than here.

For me, the first page of Etzioni's book already was sufficient evidence of his quality. He dedicates it to his four-year-old son Ethan:

I have walked and prayed for this young child an hour
And heard the sea-wind scream upon the tower,
And under the arches of the bridge, and scream
In the elms above the flooded stream;
Imagining in excited reverie
That the future years had come,
Dancing to a frenzied drum,
Out of the murderous innocence of the sea.

This is Yeats, and I must say it is better than Wordsworth.

In August 1962 I wrote a fifty-three-page report; "Implications of New Weapons Systems for Strategic Policy and Disarmament," classified secret, as my contribution to the work of the Disarmament Agency, with a covering letter to my boss Frank Long. I used the last line of the Wordsworth sonnet as epigraph to the report. In my letter to Long, I listed four new weapons systems that were then at various stages of development: (1) gigaton mines (enormous hydrogen bombs to be exploded offshore in ships or submarines), (2) Soviet antimissile systems, (3) fission-free nuclear weapons (pure fusion weapons, then studied intensively in USA and USSR, fortunately proving to be unfeasible), and (4) supersonic low-altitude missiles (the Pluto missile, then under active development at Livermore). I concluded by saying, "In each of these four areas there is need for an intensive technical study of the facts and for an imaginative grasping of the political opportunities which the new technical developments may offer. The political opportunities will mostly be lost if they are not foreseen and prepared for."

It is unlikely that my report had any effect on the work of the Disarmament Agency. The agency was too busy with short-term problems to think about long-range strategy. We were deeply engaged in preparing for two kinds of negotiations, one concerning general disarmament at the United Nations, the other concerning a test ban with the Soviet Union and the United Kingdom. The test ban negotiations required constant and urgent attention. We all knew that the test ban was real while general disarmament was mostly propaganda. Every detail of the test ban negotiations had to be discussed with the U.S. senators who would finally vote on ratification of a treaty. We used to joke that the easy part of our job was negotiating with the Russians, the hard part was negotiating with the senators. We knew that the fate of the test ban would be decided in 1963, and it was our job to get the treaty signed and ratified. Twenty years later I published the book Weapons and Hope *(1984) explaining to the public the long-range problems of war and peace that we did not have time to pursue at the Disarmament Agency.*

DECEMBER 20, 1962

There was an executive committee meeting of our Federation of American Scientists. My year as chairman is now more than half over, and I have not been energetic. I will leave the organization much the

same as I found it. My talent does not lie in the business of organizing meetings and committees. It was worthwhile to try it once, but it
is not my métier. I am most effective when I am sitting quietly alone
and writing. In future I will leave the position of chairman to others.
The Washington meeting was amiable and disposed of a mountain of
business. But I feel I am riding the federation rather than driving it.
Our main problem is still to get more members, and this problem is
still unsolved.

17

MARCHING FOR JUSTICE

THE YEAR 1963 was full of public turmoil, triumph, and tragedy. It was the year when I was most heavily involved in public affairs. The letters in this chapter are mostly describing history as I saw it unfold in Washington. President Kennedy had opened the door for me by creating the Arms Control and Disarmament Agency. During the short years of his presidency, ACDA was an independent branch of the government, intended to wage peace just as the Defense Department was intended to wage war. While Kennedy was empowering ACDA to wage peace, he was sending troops to Vietnam and involving the United States on the losing side of a civil war. Then Kennedy was assassinated, and to our surprise President Johnson became a stronger force both for good and for evil. Johnson was stronger for good when he pushed through Congress the Civil Rights Act and the Voting Rights Act, the laws that gave legal equality to black Americans. He was stronger for evil when he sent hundreds of thousands of young Americans to Vietnam to fight an unwinnable war.

Nowhere mentioned in the letters is the birth of our fifth child Miriam in September. That event is missing in the letters because my sister Alice came over from England to help Imme, and I did not write letters while Alice was with us. On the day of the birth, there was a public meeting in Princeton to organize a citizens' coalition for peace. As I was preparing to go to the meeting, Imme announced that she had to go to the hospital. Reluctantly I abandoned the meeting and drove with her to the hospital. While we were there, a thunderstorm raged outside with torrential rain. Flashes of lightning and crashes of thunder greeted Miriam when she arrived.

Miriam did not disappoint us. She remained a rebel all her life. She disliked her name and changed it to Mia as soon as she had the legal right to do so. She trained as a nurse at Columbia University and led a rebellion of the nursing students. At that time the medical students had cadavers to dissect, but the nursing students had none. Mia condemned the discrimination as unjust and unhealthy. Nurses need to know at least as much about human anatomy as doctors. Mia's rebellion succeeded, and the nursing students got their cadavers. Some years later she was ordained as a Presbyterian minister. She continues to work as a nurse as well as a pastor, taking care with equal skill of bodies and souls.

JANUARY 7, 1963

On New Year's Day we had a cozy tea party with homemade cake and candlelight. Helen Dukas came and brought along Margot Einstein, the stepdaughter of the great man. Margot was thin and frail, an old lady dominated by the forceful Helen. The four children were unusually sweet, and Margot talked more than I have ever heard her talk. She is usually shy, but this time she opened her heart and evidently enjoyed the party as much as we did.

JANUARY 16, 1963, NEW YORK

I am in New York for two days doing my job as FAS chairman. Yesterday we had three hours of the executive committee, followed by four hours of the council. Our main business was to plan a new series of educational breakfasts for congressmen and senators in Washington, to impress on them the fact that not agreeing with the Russians on some kind of disarmament may be more risky than agreeing. We authorized the Washington office to spend a thousand dollars on breakfasts. FAS membership has gone up from 2,200 to 2,450, and the council was enthusiastic about the chances of doing something effective in Washington. I have not been an outstanding chairman, but my time in office has gone by without FAS falling apart. It seems to be at least as vigorous now as it was a year ago. I had a letter from the Disarmament Agency asking me to come to Washington again for two months next summer. I said yes. I am glad that they invited me instead of waiting for me to invite myself.

We had a surprise telephone call from James Lighthill, just arrived in New York on his way to meetings. I asked him to come down to Princeton. He talked a great deal about the Nassau meeting [*in the Bahamas*] where he had taken part as scientific advisor to [*Prime Minister Harold*] Macmillan [*during his three-day meeting with President Kennedy to discuss the British purchase of Polaris missiles from the United States*]. He said he advised Macmillan to reject the American offer, but Macmillan accepted it. It was interesting to me to find that Lighthill has an unshakeable belief in the future of England as a great power. He believes that it really does some good for England to build bombs and rockets and all the rest of it. He talks about these things just like de Gaulle. I suppose he has to believe in all this in order to do his job well at Farnborough. All my arguments were unable to shake his faith. James enjoyed our household so much that he stayed overnight and slept in my pajamas.

James Lighthill was one of my closest friends since we arrived together at Winchester College at the age of twelve. We were both in love with mathematics and worked together at Winchester through the three fat volumes of Jordan's Cours d'Analyse *which James discovered in the school library. In 1943, when I went to work at RAF Bomber Command, James went to work at the Royal Aircraft Establishment at Farnborough. In 1945 I was best man at his wedding. He later became a world-class expert in fluid dynamics and a high-level advisor to the British government. His biggest contribution to science was the understanding of jet noise. He discovered the famous eighth-power law, which says that jet noise increases with the eighth power of the jet velocity. He returned to Farnborough as director of RAE, a huge organization responsible for design and development of airplanes, at the age of thirty-five. He was one of the chief promoters of the Concorde supersonic airliner. I argued with him about the Concorde. I said that the Concorde was too small and too expensive to compete in a commercial market with ordinary subsonic airliners. I said that ordinary people like us would never be able to afford to fly in Concorde. James replied, "Ah, but people who matter will be able to afford it." People like us do not matter. I knew then that I would never win the argument.*

February 5, 1963,
Manned Spacecraft Center, Houston

No, I did not volunteer to be an astronaut. I came here for three days for a meeting of the Astronomical Advisory Committee to the Space Agency. Mars is very bright now in the evening sky. If all goes well, my friend Martin Schwarzschild will launch his balloon telescope on Friday night and take the first photographs of Mars from eighty thousand feet altitude, where the unevenness of the air no longer disturbs the optical definition. If he has good luck, he will see a clearer picture of Mars than anybody has seen before.

The meetings had two main items of business, the Orbiting Astronomical Observatories (OAO) and the problem of the scientific training of astronauts. The OAO will be the first substantial pieces of scientific equipment to be put into space. The first OAO will fly about the beginning of 1965. We are pleased with the way this program is coming along. The astronauts are another story. This Manned Spacecraft Center has been from the beginning the most unscientific of the space establishments. There are no scientists here, and there is not even a good university within hundreds of miles. The emphasis is entirely on the practical problems of getting people to the moon and back. Nobody cares much about what the people may do when they are there. Space flights are considered as sporting events pure and simple.

I am quite sympathetic to the "sporting event" view of space flight. It is probably wise not to expect much in the way of serious science to come out of the manned flights for the next ten years. However we want to educate the people here, and the astronauts in particular, to be aware of the scientific jobs that they might be able to undertake. The astronauts are willing, but their time schedule makes it impossible for them to pick up more than a smattering of science. We are trying to change this. Yesterday we talked for two hours with Deke Slayton, the astronaut who was rejected for a flight last year because of some minor heart trouble. He is an intelligent and helpful person. But he was available to talk to us just because he is medically suspect. The others are too busy preparing for their flights.

The U.S. space program has always consisted of two separate parts, the manned program and the unmanned program. The Manned Spacecraft Center at Houston is the control center for manned missions. Unmanned missions are controlled from Goddard Spacecraft Center in Maryland and the Jet Propulsion Laboratory in Pasadena. Our Astronomical Advisory Committee held meetings at all three places. President Kennedy gave an enormous boost to the manned program when he announced the Apollo program to land astronauts on the moon. For science and astronomy, unmanned missions are far more effective than manned missions. But for the politicians and for the general public, the manned missions are more exciting. The main concern of our committee was to keep the unmanned program alive while the big money was flowing to the manned program. Our efforts were in the end successful. When six brilliantly executed moon landings had used up the available money, and the Apollo program was abruptly terminated, the unmanned missions continued to explore the universe.

Last Wednesday I went to Haverford College to talk about problems of disarmament. Haverford is one of the little Quaker colleges near Philadelphia. I was invited by the student peace council. I enjoyed the Haverford students especially because they reminded me of the students I knew at Cambridge in 1941–43. They were ostentatiously ill-dressed, uncombed, and unwashed. Rebellious against all generally accepted conventions, particularly against the prevalent materialistic standards and against the political establishment. This is the way students ought to be. One misses it completely among the pampered young people in Princeton.

JUNE 8, 1963

A week ago we had George's tenth birthday party, postponed because he wanted to have it out of doors. There came fifteen boys, and they were real ruffians. The main entertainment was squirting water pistols at each other until they were all completely soaked and the garden inundated. They tormented Franklin Yang and shouted "Chinese Chinese" at him until he took refuge up a tree. I was so angry that I almost sent them all home without any supper. I regret to say that George did not come to Franklin's defense. It seems that those of us who are not brutes are mostly cowards.

JULY 9, 1963, WASHINGTON

[*On July 4*] we walked out in front of the White House and watched the fireworks. After the official fireworks were over, there were large numbers of children, all negro, running around on the grass and letting off their own little fireworks which they had brought with them. Dorothy and Emily stood fascinated watching the children play. Three times in a few minutes, one of the negro children came up to Dorothy to give her a firework which she could hold in her hand. I was very much touched by this.

The agency is in a hectic state preparing for the Moscow negotiations which are due to start in a few days. Our group of people went to see Kennedy today and will leave for Moscow on Thursday. I do not have any clear idea of how much freedom of negotiation they will have. Probably very little. I expect nothing very significant to come out of these negotiations, but I may be wrong.

AUGUST 5, 1963,
DISARMAMENT AGENCY, WASHINGTON

I am in the middle of three separate campaigns: (1) the preparation of papers inside the government for the Senate debate on the Test Ban Treaty ratification, (2) a public campaign to bring two scientists from every one of the fifty states to Washington, where they will talk to their respective senators in defense of the treaty, and (3) a campaign going on inside and outside the government against the present policies in Vietnam. In addition to these various activities I try to get home for weekends.

(1) The official work at the agency has become more directed and everybody is more dedicated, now that there is a treaty to fight for. Everybody understands that this fight will be decisive for us. If the treaty is defeated, then the Disarmament Agency will be dead. Kennedy will be impotent, and there is no chance that any negotiations can be carried on with the Russians on any subject at all. Whether or not we particularly like this treaty, we are all fighting hard for it. I myself had my little moment of glory while the negotiations were in progress in Moscow. There was a sticky point which held up the agreement for several days. Averell Harriman cabled back to Washington, "May we give way on this one?" Kennedy picked up his telephone and called William Foster, the head of this agency, and asked him what he thought about it. Foster said, "I must consult with my staff about this." So Foster called Al Wadman and asked for his advice. Wadman said, "I don't really know much about this, but maybe Dyson would know." So Wadman came over to me and talked it over. I had some expert knowledge because of the time I have spent designing weapons at Livermore. I said to Wadman, "This is ridiculous. Of course they should give way." So Wadman called back Foster, and Foster called back Kennedy, and the cable went back to Moscow, and the treaty was signed.

Frank Long came back later from Moscow and gave us a complete blow-by-blow account of the negotiations. He said the bargaining was very hard on both sides, and they were not at all sure until the last day that they would be able to agree on a treaty. So it seems that my little contribution helped. This means the death blow for Project Orion. I am really sorry about this. But I had to admit in my own mind that no single project of that sort could be allowed to stand in the way of a treaty. I spent an evening with Ted Taylor here in Washington the day after the treaty was signed. He was very upset, having given five years of his life and energies to fighting for Orion. He took it bravely, even when I told him I was the Judas who had betrayed him to the enemy. He says he plans to wind up Orion in the next six months and then take a year in Europe to get away from it all.

(2) The public hundred-scientist campaign is starting off well. We had great sport trying to dig up scientists from cultural deserts like Mississippi and Alabama. Unfortunately it is precisely the places like Mississippi and Alabama which have the most crucial senators. The gathering in Washington will be on August 16 so time is short. The FAS already gave a press conference at which I was on the platform with two others. About fifty reporters came to fire questions at us. It was surprisingly easy to answer the questions, and I think we did well. Unfortunately the newsmen who were there were all in favor of the treaty already, so we were preaching to the converted. The senators will not all be so friendly. This week I am supposed to appear at the Press Club in a public debate against Edward Teller. That will be more of a challenge.

(3) We are involved with Vietnam for several reasons. Yesterday and today I spent interviewing a German doctor called Wulff from Freiburg, who has just come home from two years of doctoring in Vietnam. He had a wealth of first-hand information about misdeeds of Vietnamese and American troops. We plan to fight inside the government for a change in policy, and then if that fails we will do what we can to arouse the public by means of FAS and similar organizations. On August 28 will be the grand negro march on Washington. So it will be an exciting summer.

AUGUST 19, 1963

There is a great satisfaction in feeling the strongly expressed approval of mankind for what we are doing. While we work downstairs in the State Department building, there is a steady stream of ambassadors and other important people upstairs coming to sign the treaty. The number of countries who have signed is now sixty plus. At the same time, I do not feel that the distinction between peace-mongering and war-mongering is nearly as sharp and clear as most people suppose. We divide ourselves jocularly into peace-mongers and war-mongers. Roughly speaking, the peace-mongers are those who are willing to accept some military disadvantages in order to come to political settlements which may make the world more stable. The war-mongers consider that we have a better chance of reaching stable political settlements if we maintain the maximum military strength. Both sides are sincerely striving for peace, and even the disagreement about methods is not very sharp.

The Test Ban Treaty is supported by most of the people who are usually classed as war-mongers. Among my friends only Edward Teller is opposing it strongly. I admired his courage last Wednesday when he came to the Disarmament Agency and argued for two hours with a roomful of people, not one of whom was disposed to agree with him. He gave a brilliant performance, which earned the respect of even those people who hate him bitterly. His strongest argument against the treaty is that it says we shall not use nuclear explosions even in self-defense in time of war. It is in fact true that that is what it says. However, Dean Rusk and other high-up people have said it says nothing about use of bombs in war. Rusk said explicitly that the United States will maintain freedom of action to use nuclear explosions in war whenever our vital interests require it. On this point, Teller made a most eloquent speech describing the German invasion of Belgium in 1914 and explaining how this was justified to him as a schoolboy in Hungary with the excuse that every country must be allowed freedom of action to break treaties if the supreme interests of the country were threatened. I support the treaty because I am in favor of a No First Use pledge for nuclear weapons anyway. But I must admit that for the average citizen who does not believe in No

First Use, the argument of Teller ought to cause some anxiety. The treaty is at least ambiguous enough to create a strong political pressure towards No First Use.

August 27, 1963

Today was my great day. I made my speech before the Senate Foreign Relations Committee, on behalf of FAS, supporting the Test Ban Treaty. The speech seems to have been well received. Many people congratulated me afterwards. I had taken care not to show it to anybody beforehand so they would not try to alter it. When several people try to rewrite a speech it always ends up as a mess. This one was at least my own. After the speech the senators asked some questions, mostly about the Pugwash meeting I had been to in Cambridge last summer. Their questions were sensible and not designed to be harassing. I told them about the Pugwash meetings without feeling on the defensive. The chairman of the Foreign Relations Committee is Senator William Fulbright of Arkansas, and he is not stupid.

Pugwash meetings are international meetings of scientists discussing political problems. They meet as private citizens and not as representatives of governments. The meeting at Cambridge (England) in August 1962 was a big one, with twenty people each from the United States, the USSR, and the UK, and twenty more from other countries. Nobody came from the People's Republic of China. The main subject of discussion was the formulation of disarmament agreements that might be acceptable to the governments of the leading countries. For me, the formal sessions were less interesting than the informal encounters with other participants. The two most impressive people that I got to know were Joseph Rotblat, the secretary-general of Pugwash, and Vladimir Pavlichenko, the KGB official who came with the Soviet scientists to keep them in line. Rotblat was unique among the scientists at Los Alamos as the one who resigned and walked out in 1944 when it became clear that the German nuclear bomb project had failed. Pavlichenko was unique

among the Russians as the one who had good communication with his own govern-
ment and was a personal friend of Nikita Khrushchev.

The hearings are much more comfortable than I had expected.
They are held in a magnificent hall with marble columns and painted
ceiling. The witness sits at a small table in the middle of the room,
and the committee is arranged in a semicircle around him. Just before
me came George Meany, the president of the American Federation
of Labor, a huge man resembling Ernest Bevin in size and also in
character. He announced his support for the treaty and then launched
onto a ferocious diatribe against the Russians and all their works.
Tomorrow the great negro march on Washington will be here. These
are stirring days.

AUGUST 28, 1963

I would like to write to you about today's events while they are
fresh in my mind. I stopped work at eleven a.m. and walked from the
State Department down to Constitution Avenue a few blocks away.
The broad avenue was completely cleared of traffic and full of people.
I walked to the end of the avenue where the march was beginning,
then joined the people and walked with them to the other end where
they arrived at the Lincoln Memorial. I found it profoundly moving
to be marching with all these people. Just for once to be seeing history
made instead of only reading and writing about it. The march was
quiet. No music and no stamping of feet. Just people strolling down
the avenue in their own time.

Each group of people carried banners saying where they came
from. Occasionally there would be some shouting and cheering when
a group came by from one of the really tough places, Birmingham,
Alabama, or Albany, Georgia, or Prince Edward County, Virginia.
Most of these people from the Deep South had never been away from
their homes before, and they had never had anybody to cheer for

them. It must have been quite an experience for them to see so many friends all together. There was an interesting difference between the northern and the southern people. The northerners were mostly rather well-dressed family people, husbands and wives, many of them union members who were brought to Washington by their unions. The southerners were much younger, most of them hardly more than children, and they looked like the hope of the future with their bright faces and gay clothes. Of course, in the bad southern states a man with family responsibilities cannot afford to be put in jail, so it is necessarily the young who must carry on the fight. Some of these southern groups sang their freedom songs very beautifully while the northerners listened.

The march along Constitution Avenue started at eleven and went on for three hours without stopping. They said 250,000 people came. I would guess they were about 80 percent northern negro, 10 percent northern white, 10 percent southern negro. The weather was kind, a temperature of eighty-two with cool breeze and sunny sky. All the people there were enjoying the outing. They had been told to leave their children at home, and that made it different from an ordinary holiday crowd. It was quite relaxed, but very serious in the underlying mood.

From two till four they had the official speeches at the Lincoln Memorial. It was very effective to have the huge figure of Lincoln towering over the speakers. The speeches were in general magnificent. All the famous negro leaders spoke, except James Farmer, who sent a message in writing from a Louisiana jail. The finest of them was Martin Luther King, who talks like an Old Testament prophet. He held the whole 250,000 spellbound with his biblical oratory. I felt I would be ready to go to jail for him anytime.

I think this whole affair has been enormously successful. All these 250,000 people behaved with perfect good temper and discipline all day long. And they have made it unmistakeably clear that if their demands are not promptly met, they will return one day in a very different temper. Seeing all this, I found it hard to keep the tears from running out of my eyes.

SEPTEMBER 25, 1963

In today's newspaper comes the news that the treaty is ratified as expected. So that struggle now passes into history.

NOVEMBER 14, 1963

We had a surprise call the other day from my old friend Oscar Hahn. He telephoned to say he would come and see us in Princeton, so we invited him for lunch. He talked about his uncle Kurt who is now back in Germany at the Atlantic College, his latest educational venture. This is an international school close by, but not identical, with the first school which he founded at Salem in South Baden. Oscar said the original Salem school and the Gordonstoun school are both flourishing, but Kurt is not interested in anything once it is established and running smoothly. The new Atlantic College is only two years old and is still keeping him happy. Oscar invited me to go over and visit Salem. It is quite near to Zürich, being just across the frontier opposite to Lake Constance.

NOVEMBER 28, 1963

Shortly after the Kennedy assassination was announced, I happened to be at the post office sending off a Christmas parcel. In front of me in the line was a negro, and he started to talk with the clerk behind the counter. "Well," he said, "there's some people on this earth who'd shoot at Jesus Christ, if he ever came back here. The trouble is all this ideology and hatred that people are taught. In my family I don't allow the kids to talk hate against anybody. I won't have any of that sort of talk in my house." After this little speech, the whole crowded post office stood silent in quiet admiration.

You say that you felt disappointed. I suppose you mean that you had expected the United States to behave in a more civilized fashion. For me, the reaction to Kennedy's killing was different. I do not feel disappointed. It is a great pity that Kennedy is dead. But to me the moral shock of his killing was much less than that of the killing of Medgar Evers, the negro who was killed in exactly the same way in Mississippi last summer. Evers was an even braver man than Kennedy and is probably harder to replace.

I suppose I have expected that the next year or two in this country will be bloody and brutal. It has seemed impossible to imagine the country taking serious actions to deal with the negro problem, unless forced to it by a major outbreak of violence and bloodshed. I have expected that there would be violence and bloodshed, probably of a peculiarly uncivilized kind, before the end of next year. In Washington last summer I have heard one of Bobby Kennedy's staff, who works on the civil rights cases, say the same thing in surprisingly unqualified language. In this situation the murder of Kennedy comes not as a disappointment but rather as a natural consequence of the general unrest. It seems that Lee Oswald was not himself concerned with the negro problem at all. Still he certainly had heard of the murder of Evers, and he had lived in this atmosphere of bitterness and violence which has always been particularly bad in Texas. So I consider it not pure accident that Kennedy should have been killed just now, in the middle of the racial crisis.

What I hope and even believe is that the death of Kennedy may have caused a sufficiently strong revulsion against violence, so that a bloodbath of Negros can be avoided. I was especially hopeful after Johnson's speech in Congress yesterday. Possibly the murder of a president can change the atmosphere enough, so that the negro demands can be accepted peacefully.

DECEMBER 12, 1963

I saw Oppenheimer the day after he was at the White House [*to receive the Enrico Fermi Prize, an official reward for public service, and a tacit apology for the denial of his security clearance in 1954. The award had been planned before the assassination, to be given by Kennedy, and then approved a second time to be given by Johnson.*] He was enthusiastic about Johnson. He said he had been invited for a private session with Johnson before the official presentation ceremony and that Johnson talked "exactly right" about the important things. He would not be more specific than this. I presume this means that Johnson showed he takes seriously the long-range problems of war and peace and not only the immediate issues. At the ceremony Oppenheimer had his wife Kitty and both his children present to witness his triumph. He made a very apt little speech which you probably have read. He said, "It is possible that you, Mr. President, required some charity and some courage in order to make the award." Afterwards Oppenheimer saw Mrs. Kennedy and told her that originally the speech had said, "some charity, some humour, and some courage," but that when Kennedy died, Kitty had insisted that the humour should be left out. Mrs. Kennedy then said, "Yes, you were right to put it in, and Kitty was right to take it out." It seems this story gives an accurate picture of the difference between Kennedy and Johnson. But there is no doubt that Johnson is far better than most of us had expected.

This Sunday I am flying to Dallas for a three-day meeting on astronomy and gravitation. The meeting is held at a new institute called the Southwest Center for Advanced Studies, intended to put Texas on the intellectual map. It is exceedingly bad luck that Dallas has now such an odious reputation. What made it especially bad luck was that Kennedy had prepared a speech singing the praises of the Southwest Center for Advanced Studies. This was the speech he would have delivered in Dallas if he had not been killed.

DECEMBER 18, 1963

The meeting was a grand affair. Nobody stayed home because it was in Dallas. The meetings were good because we had something to talk about. The subject was the new astronomical objects which have been discovered during the last year [*later known as quasars*]. Nine are now known, and more are being found every month. The nearest one is 1,600 million light-years away, and still it is so bright that they have several thousand photographs of it going all the way back to 1870. These things are totally different from what anybody had expected. They must be understood if we are to pretend to understand anything about the general plan and history of the universe. The observers swept us all into silence with the solidity and clarity of their observations. Afterwards the theoreticians and physicists mumbled pathetic fragments of ideas, not having anything coherent to say. I kept my mouth shut.

I enjoyed a long evening talking with a young man called Yuval Ne'eman from Israel. He made his reputation two years ago with a brilliant theory of particles. He was then a student at Imperial College in Kensington. The remarkable thing about Ne'eman is that he is not really a physicist but a professional soldier. When he was still a boy, he fought in the War of Independence of 1948 and ended up as a colonel. He then quickly became deputy chief of intelligence on the Israeli General Staff. During the Suez campaign of 1956 he was flown to the British and French headquarters on Cyprus in a fruitless attempt to establish some coordination between the Israeli and British-French armies. Afterwards he was military attaché in London and bought two submarines for the Israeli navy. It was during the period when he was buying the submarines that he found time to study physics at Imperial College. I think it is an unparalleled achievement in modern times for a man who had been a professional soldier for ten years to make a major contribution to physics. I found his stories about the military world illuminating. He had been through the French Staff College, and in London he had got to know well most of the leading generals and admirals.

Ne'eman continued for many years to perform brilliantly in his triple career as physicist, soldier, and politician. He remarked to me once that the greatest joy he had ever experienced was to command a brigade of tanks in battle. After the War of 1967, when Israel defeated the combined armies of Egypt, Jordan, and Syria,

Ne'eman proposed to give the populations of all the occupied territories Israeli citizenship. He said this was the great opportunity for making peace in the Middle East, when the shock of defeat would allow the Arab populations to accept annexation into Israel, but the acceptance could work only if they were given full equality and civil rights as Israeli citizens. Unfortunately that opportunity was missed, and we will never know whether it would have succeeded.

Ne'eman was later known to the public as a right-wing politician who vehemently opposed the Camp David Accords of 1978 between Israel and Egypt. The accords returned to Egypt the entire Sinai peninsula occupied by Israel in the War of 1967. Ne'eman had a vision of the future with Greater Israel, including the Sinai and the West Bank and the Golan Heights, uniting in a single country the old Israel of 1948 and all the occupied territories. Ne'eman saw himself as a liberal humanist, giving generously to the Arab populations all the rights and privileges of Israeli citizenship. His vision might have been realized in 1967, but by 1978 it was certainly too late. Ne'eman continued until his death in 2006 to be a right-wing politician, with an unexpectedly generous grasp of the Arab point of view.

18

SITTING IN JUDGMENT

THE AMERICAN LEGAL SYSTEM uses two kinds of juries, the petit jury of twelve citizens who sit through criminal trials and decide the guilt or innocence of the accused, and the grand jury of twenty-three citizens who conduct quick preliminary hearings to decide whether accusations should go to trial. I had several times been summoned to serve on a petit jury in Trenton, but each time I was dismissed because the lawyers do not like to have college professors as jurors. Decisions of a petit jury must be unanimous, and college professors are very rarely unanimous. In the fall of 1963 I was summoned to serve on a grand jury, where decisions are by majority vote. The lawyers had no objection to me, and I agreed to serve, beginning in January 1964. In the same month there was a meeting in Boston to make plans for the exploration of Mars.

JANUARY 13, 1964

We had three days of Mars meetings. It was a splendid gathering. The two most famous observers of Mars from the French observatory at the Pic du Midi were there; also various astronomers, space experimenters, biologists, and chemists. About twenty-five people altogether. The purpose of the meeting was to advise the Space Agency about the the future of the program for observing Mars, especially to design sensible experiments to find out whether there is life there or not. The two Frenchmen dominated the discussion, since they are the only ones who know their way around on Mars. In three days we all became expert in referring to the regions of Mars by their beautiful classical names, running easily from Syrtis Major through Nepenthes-Thoth and Mare Cimmerium to Amazonis, Arabia, and Hellas.

The evidence for life is still as tantalizingly inconclusive as it always was. But we are going to know the answer in a few years for sure. The first rocket probe will go to Mars this year and make preliminary measurements. An instrument package should get down to the surface by 1969 and tell us in detail what goes on there. Meanwhile facts are slowly accumulating and theories are growing firmer. The biologists have done experiments which prove that many terrestrial bugs can survive the Mars environment.

Fifty years later, we still do not know for sure whether there is life on Mars. Large numbers of orbiting instruments have surveyed the planet, and several roving observatories on the ground have explored the surface, without finding any evidence of life. It is possible that life exists or has existed at places below the surface where rocks are warm and water is liquid.

The grand jury turns out to be fascinating. I go every Tuesday until the end of March. Assuming I can fight my way through the snowdrifts, I will be there tomorrow morning. Last Tuesday we had our first session. We had three larcenies, one assault and battery, one forgery, and one contribution to the delinquency of a minor. We sent five of them to trial and dismissed one. It is fast and sketchy work. The purpose is to stop the police from bringing anybody to trial without a reasonably convincing case. We hear only the prosecution and one or two witnesses for each case. After the evidence, we discuss the case for a few minutes among ourselves and then vote.

One comes away from these cases with admiration for these people, criminals and victims alike, for their ability to live at all under the conditions they describe. Mostly they are negroes and Puerto Ricans.

Such a tremendous gulf between their conditions of life and the lives of the prosperous citizens who sit on the grand jury. I cannot blame them if they have little respect for us. It is difficult for us to deal with them fairly, because the mere fact of their being homeless drifters in the back streets of Trenton makes it too easy to believe them guilty. The most doubtful case was a man accused of breaking and entering a house, the only evidence against him being the word of an accomplice without any material proof. I felt that nobody should be convicted of a crime on this evidence alone and therefore we ought to dismiss the case. But the vote was to send him to trial. It is a good education in legal principles, and I am glad I was chosen to take part.

Service on the grand jury is a serious commitment. One full day of sessions every week for three months. I was lucky to be able to spare so much time. The jury hears the case for the prosecution and decides either to dismiss the case or to let it go to trial. The proceedings are secret, so if the case is dismissed, the public never gets to hear about it. The defendant needs to prepare a defense only if the case goes to trial. Our meetings were in the Trenton Courthouse. There was no break for lunch, so our decisions tended to become quicker and harsher as we grew hungrier at the end of the day.

MARCH 27, 1964

Imme and I went with Emily to buy a birthday present for George [*George then aged eleven, Emily aged two years and eight months*]. We bought him a plain wooden chest which we hope he will use for storing his belongings. When Emily saw it, she said, "When Kennedy was shot he had a box like that."

A case of incest at the Trenton Court. A father accused by his thirteen-year-old daughter of having forced her to bed with him for three years. The girl, being roughly questioned by the prosecutor, broke down in floods of tears and admitted her whole story was false. In this case the grand jury justified its existence in protecting the man from a trial which would have ruined him even if he had been acquitted. Next question, what does one now do with the girl? Luckily this is not our responsibility.

Our decision to dismiss the incest case was not so obviously right as I implied in the letter. Our jury consisted of eighteen men and five women. The vote in the incest case was eighteen to five. The men all supported the father and the women all supported the girl. If the jury had been five men and eighteen women, the vote would undoubtedly have sent the case to trial. To do justice in such cases is not easy. The mother of the girl was mentally ill and confined to an institution. The father had to deal alone with a rebellious teenager. The girl wanted the freedom to live her own life. After fifty years, I am still not sure whether what we did was justice. The judge who presided over the grand jury sessions shared my doubts about the justice of the jury system. After the last of our sessions ended, he invited us all for a farewell dinner. At the end of the dinner he stood up to make a short speech. "Before we say good-bye," he said, "I would like to give you some useful advice. If you are ever in trouble with the law, you have a choice, to be tried by a judge or a jury. If you are innocent, choose a judge. If you are guilty, choose a jury."

APRIL 25, 1964

Abdus Salam, a former student of mine, is an exceptionally able young man. After doing a few years as a theoretical physicist at Cambridge and Princeton, he decided to return to his native Pakistan to work for the good of his people. In Pakistan he was a professor of physics and head of the Atomic Energy Commission, but he could not do anything effective to help the country from the inside. After a few years he left Pakistan to take the chair of theoretical physics at Imperial College. Now from his independent stand in London, he has become a major international figure. He has an excellent department in London, he is official scientific adviser to the president of Pakistan, and he has founded an Institute of Physics with United Nations money in Trieste. The Trieste institute is similar to this institute at Princeton but more easily accessible to people from the poorer countries, especially Eastern Europe, the Middle East, and Pakistan. Politically, it has been a great success, and Salam has pushed it through the United Nations in masterly fashion. I believe it will also be scientifically successful. An ambitious young man with a chair at Imperial College can have the world at his feet.

The main thing we have done recently was to give Oppenheimer a sixtieth birthday party. I sat next to Kitty Oppenheimer. She was in unusually good spirits. The main ceremony at the party was the presentation of the special issue of the *Reviews of Modern Physics* dedicated to Oppenheimer, on which I had been doing most of the editorial donkey work for the last year. The first copy was rushed down from New York hot from the press the same day. Oppenheimer seemed to be genuinely surprised and moved.

The letter does not mention the agonizing difficulties that I encountered in putting together a respectable collection of chapters for the Oppenheimer volume. I began by writing letters to all the people who had worked with Oppenheimer, especially in the early years when he was most active as a scientist. Very few responded positively. Most of them made polite excuses for saying no. The saddest reply came from Max Born, one of the bunch of geniuses who invented quantum mechanics in the 1920s. Born had been Oppenheimer's adviser when he was a student in Göttingen and collaborated with him on the quantum theory of molecules. Born wrote a bitter letter, complaining because Oppenheimer was ungrateful for his help and had never invited him to Princeton. The unfortunate fact was that Oppenheimer had a difficult temperament and made more enemies than friends. I had to write many more invitations before I had enough articles to make a presentable volume.

Oppenheimer's shifting moods cost him many friendships. I remained his friend because I did not take his temper tantrums personally. In spite of the temper tantrums and shifting moods, he was a loyal public servant and a wise observer of the human scene.

May 1, 1964

A bundle of telegrams, one from Kitty and Robert Oppenheimer, said, "Congratulations on your election to the National Academy of

Sciences." Elected at the same time were Tsung-Dao Lee and several other good friends. This honor is the American equivalent of an FRS [Fellowship of the Royal Society]. Like the Royal Society, the National Academy covers all areas of science. It has about nine hundred members compared with the Royal Society's six hundred. I am now both FRS and MNAS, a rare combination. Apart from honor and glory, being an academy member gives me an official standing which is helpful if I am sitting on committees or advising the government. I now belong to the Establishment.

Imme and I were invited in 2012 by the Royal Society in London for an official celebration of my sixtieth anniversary as a fellow. Paul Nurse, then the president of the society, gave an open-air lunch party on the roof of the Royal Society building, a magnificent viewpoint overlooking the Horse Guards parade and the historic monuments of central London. It turned out that Queen Elizabeth and I were the only people who had survived sixty years as fellows. The queen was having her annual Garden Party at Buckingham Palace the same day, so she did not come to my party and I did not go to hers.

MAY 23, 1964

I recently read three books which I found interesting. One is *Profiles of the Future* by Arthur Clarke [1962], an English science fiction writer. He has written many books, and all of them are good. Another book I read was *The Analytical Engine*, a popular book about computing machines by Jeremy Bernstein [1964]. The third was brought to our house by Esther from the public library. It is *Face to Face*, the autobiography of a blind young man who came from India to the United States and became a successful writer [1957]. His name is Ved Mehta.

Imme and I, for the first time in years, went to a cocktail party at the home of some friends of ours in New York. We first met Jeremy Bernstein, who I knew would be there, and talked to him about his book. Then I was introduced to a lively middle-aged gentleman who turned out to be Arthur Clarke. I spent an hour monopolizing Arthur Clarke and exchanging ideas with him about space and the universe. While I was talking to him, I saw a blind Indian come in at the door, and Imme said, "I bet you that is Ved Mehta." I said, "No, that is absurd." But as usual Imme was right. I spent another happy hour talking with Ved Mehta. I will never believe any more that improbable things don't happen.

This unlikely coincidence brought me two lasting friendships. Arthur Clarke and Ved Mehta continued for many years to write wonderful books. Clarke's Childhood's End *(1953) and Mehta's* Walking the Indian Streets *(1960) are for me the most memorable.*

In 1964 I was invited to spend a year at the University of California San Diego, a brand-new and rapidly growing university with a campus adjoining the General Atomic buildings where I had worked six years earlier. In the fall, Imme's parents came for a month to live with us in California. During that year, I made my last attempt at research in particle physics. In collaboration with the Vietnamese physicist Nguyen Huu Xuong, I worked out a theory of particles based on the group SU6, which gives the particles a six-fold symmetry. We made some bold predictions that could easily be tested by experiments. We predicted some new particles that would have confirmed our theory in a spectacular fashion if they existed. The tests were done, and the particles do not exist. Nature does not dance to our tune. Xuong and I agreed to give up particle physics. I switched my attention to astrophysics, and he switched to a successful career as a biologist.

During the year at UCSD I frequently visited the Salk Institute for Biological Studies. There I got to know the chemist Leslie Orgel who taught me everything I know about the origin of life. He spent his time inventing theories of the origin of life and doing chemical experiments to test his theories. The experiments never gave him a clear answer. As a result of the friendship with Orgel, I continued to think and write about the origin of life, one of the great mysteries to which we owe our existence.

OCTOBER 15, 1964

The first news of Papa's death came to us from George. He was alone in the house, and there came a call from the *New York Herald Tribune* asking for information for Papa's obituary notice. George gave them the information and then came over to my office to give me the news. That evening after supper Edwin Jung [*Imme's father*] talked about his experiences of death. In his work he has seen much of it. He said he is rarely depressed by the death of old people. They usually are able to die easily and peacefully. He is only scared when he has to deal with young people dying, especially the young soldiers during the war.

I visited my father in Winchester for a week, shortly before he died. We knew it would be our last good-bye. Each morning I listened while he played Bach and Haydn on the piano. He could not play modern music anymore, but Bach and Haydn remained with him to the end.

The letters you have had from Papa's friends must be a great comfort. It does not surprise me that they remember him primarily for his charm and vitality at meetings and rehearsals. This is what I would have expected. After all, it was for this that they made him director of the Royal College of Music and chairman of the Carnegie Trust. Just being a first-rate musician would not have got him into those positions. What does surprise me is that anyone in this hectic life should find time and energy to sit down and write you a letter. This I consider a real marvel and a clear indication of Papa's greatness.

My father died on September 29, 1964, at the age of eighty-two. He had been declining with arteriosclerosis for some months and was afraid of becoming incapacitated. Then he caught pneumonia, and his doctor who knew him well made the decision not to give him antibiotics. He died quickly and peacefully, as he would have wished. A full biography by Paul Spicer (2014) was published fifty years later.

MARCH 1, 1965

The annual Glider Club meet takes place at the airstrip on the clifftop only a mile from our house. It is a lovely sight, ten or twelve of the big birds in the air at once and weaving patterns in the sky. This year there were a scattering of European designs, one from England, one from Poland, and two from Germany. The European gliders are made by hand with loving care, whereas the American are made in factories. Still it was a prosaic American bird which won the grand trophy. The most dramatic of the competitions was the spot-landing contest. The gliders were required to land on the bumpy airstrip with the wind behind them, so that the front tip of the glider came to rest as close as possible to a marked spot. The competition was won with a distance of one inch. Several others came within a foot. It was good to see the old glider club still flourishing and hanging on to its precarious perch on the clifftop.

I am writing a post-mortem description of the Orion Project, the spaceship project which I helped to launch in 1958. It officially ceases to exist today, after a life of seven years. Political difficulties of all kinds killed it. I am rescuing it from oblivion by writing an account of what happened and why, to be published in some magazine of wide circulation.

My account of Orion was published with the title "Death of a Project" (1965). I described how it happened that a powerful new technology, with immense possibilities for expanding human exploration of the universe, was abandoned for political reasons. My purpose was not to revive Orion but to make the public aware of the costs of saying no to a bold new dream. The reasons for killing Orion were valid and compelling, but the costs were real too.

MARCH 7, 1965

We had a treat this week, a reading of his own poems by Cecil Day Lewis. I have always been an admirer of Lewis, and now more than ever. What was impressive about this hour of reading was that I had heard none of it before. Every bit of it was new and just as good as his old stuff.

APRIL 30, 1965

Early in the week I heard that Oppenheimer decided to retire as director of the institute. This came as a surprise. He had given no hint of it when I saw him in December. Frank Yang is also leaving to be Einstein Professor at a new university which New York State is building on Long Island. This also is sudden and mysterious. I shall have to think carefully about my own position. If the physics group at the institute is to stay alive, I will have to take more direct responsibility for running it.

MAY 17, 1965

Leo Szilard was a friend of mine and also one of the great men of the age. He was the man who originally persuaded Einstein to write his famous letter to Roosevelt to start the atomic energy project. He was also the first to begin agitating seriously about the political consequences of nuclear weapons, during the war years when such questions seemed far in the future. He died a year ago here in La Jolla. It happens that Imme and I have made a close friendship with Szilard's widow who is still living here. She is a lady of sixty who is left alone in the world and loves our company and our children. She is in the same situation as Helen Dukas, dedicated to the memory of Szilard as Helen is to Einstein. Trude Szilard is a medical doctor specializing in public health and has a job in Los Angeles, so we see her only at weekends. For many years she was a public health administrator in New York. In these last weeks I begin to feel sorry to leave this place. Getting to know Trude Szilard has made a big difference. But we have to pack up our stuff and go.

I met Leo Szilard for the first time at Pugwash meetings, where he was always one of the most active participants. He liked to talk directly with the Russians and espe-

cially with the KGB man Pavlichenko, offering concrete proposals that might later reappear in official government negotiations. I also met him in Washington, where he founded an organization known as the Council for a Liveable World, giving financial support to politicians committed to peaceful solutions of international problems. Following Szilard's advice, I gave a substantial contribution to the election campaign of a young senator from Delaware, not yet famous. His name was Joe Biden.

JUNE 17, 1965

Our trip home was fast and easy. We drove it in six and a half days. The children were glad when we reached home. The reaction of Miriam was unexpected. We did not know how much Miriam understood of the meaning of this long trip, or whether she even remembered Princeton at all. When we arrived at the house, she rushed up the stairs and into her old bedroom and then dropped down onto all fours and crawled around the room with the most blissful expression on her face. Evidently she knew where she was. She had learned to walk in La Jolla, so she had never walked in her Princeton room. The next morning she again crawled around for a few minutes before she remembered to stand up and walk.

JUNE 22, 1965

We let Dorothy baby-sit for the first time. I left the telephone number 924-3637 with an arrow over it. Dorothy said, "Why did you draw an arrow?" I said, "So you can be sure which end to begin. I always used to draw an arrow for Esthi when she was baby-sitting when she was your age." Dorothy replied, "But you don't need to do it for me. I can see that if I started from the other end the four would be the wrong way round."

OCTOBER 23, 1965

We are all excited because my three friends Tomonaga, Schwinger, and Feynman won the Nobel Prize. You may remember that it was just after their great work in 1947 that I started my career by carrying further what they had begun. I am happy that the prize is given to the three of them equally. To some extent I can take credit for this, since Schwinger originally had all the limelight and Tomonaga and Feynman were struggling in obscurity. It was my big paper "The Radiation Theories of Tomonaga, Schwinger and Feynman" that first did justice to all three of them. I am now writing the historical account of their work which will appear next week.

NOVEMBER 19, 1965

I called up Daniel Singer, the Washington lawyer who works for the Federation of American Scientists. The call lasted over an hour. Singer had just come back from a month in Mississippi where he was working as a volunteer, defending negroes in civil rights cases. None of the local lawyers down there will work for negroes. I admire Singer very much. He has a wife and four small children, and he left them for a month to work without pay and with a considerable risk that he might get shot. He talked for an hour without stopping. He says things there are much worse than even he had imagined. Negroes systematically terrorized and the white people totally uninterested in any kind of reform. There is now a permanent organization of lawyers who take it in turns to go down for a month and get negroes out of jail. Five or six of them are in Mississippi each month, and they are able to do quite a lot, but they get no help from the local white people.

Daniel's wife Maxine Singer is now much more famous than Daniel. She became a leading microbiologist and organized the international meeting at Asilomar in 1975 to establish guidelines for the regulation of experiments in genetic engineering. She ended her career as president of the Carnegie Corporation, in charge of research programs in astronomy and geophysics as well as biology. Maxine and Daniel are both heroes. I am lucky to have been a friend to both of them. Their four daughters have done well in various professions. Amy, the oldest, is a professor of history at the University of Tel Aviv and an expert on the Ottoman Empire. In 2015 she was at Princeton as a member of the School of Historical Studies of the institute.

19

TWO DEATHS
AND TWO DEPARTURES

In 1966 Robert Oppenheimer and Imme's father, Dr. Edwin Jung, were dying of cancer at the same time. Dr. Jung was sixty-one; Oppenheimer was sixty-two. Both had been looking forward to twenty years of retirement, Dr. Jung with his son Diether gradually taking over his medical practice, Oppenheimer with the freedom of a former institute director. The letters begin with the appointment of the new director.

FEBRUARY 17, 1966

A pair of institute trustees came to visit me in a secretive manner and revealed the name of our new director, a man from Harvard called Carl Kaysen. They came to hear any objections I might have before the appointment would be made official. I could honestly say that I am enthusiastic about Kaysen. He is a kindred spirit. He is an economist by trade, but he became involved in military and political problems in the same way I did. While I was receiving my education at Bomber Command from 1943 to 1945, he was just six miles away at the headquarters of the American bomber command at High Wycombe, doing the same job and receiving the same education. We might even have met on the few occasions when I visited the American HQ. He is three years older than me. After the war he went back to university life but continued writing papers on strategy, bombing, disarmament, etc. Once these problems get a grip on a man, they do not let him go.

When Kennedy became president, he immediately put Kaysen into a high position in the White House. While he was there, he did a great deal to push disarmament and liberalize the government's foreign policy. While I was at the Disarmament Agency in summer 1963, Kaysen went to Moscow with the team negotiating the Test Ban Treaty. Everybody I knew in the government had a great respect for him. I will enjoy to have him here and talk politics with him. I imagine he will find it easy to run the institute with a quarter of his time and keep the rest for government affairs.

After the trustees came, I flew to Florida and spent a weekend with the Space Committee looking at big rockets. The big rockets are now going to start flights leading to the lunar landing. The latest Russian success adds excitement to the race. If all goes according to plan, we shall have some men on the moon before the end of 1968. The main purpose of our meeting was to see what scientific experiments can still be in time to go on these flights.

Yesterday came the worst bombshell. Oppenheimer has a throat cancer and is in New York having radiation treatment. The doctors say it is a superficial thing, discovered early and with a good chance of being cured. I do not know how much of this to believe.

FEBRUARY 28, 1966

I had a great piece of luck. With a young man called Andrew Lenard, I have been in the throes of creation for the last two weeks, and we solved a quite important problem. It is the best thing I have done since the summer of 1961. Now the work is done, but there will be the job of writing it up for publication which I shall also enjoy.

Andrew Lenard came to the institute as a visiting member from the University of Indiana. He started thinking about the problem which we call "stability of mat-

ter." The problem is to understand by strict mathematical proof why ordinary mat-
ter consisting of positively charged nuclei and negatively charged electrons does not
collapse into a state of lower energy. Lenard made some progress with this problem
and then came to me for help. We worked together on it for half a year before we
had it solved. Our proof was extraordinarily long and complicated. A much shorter
and more illuminating proof was found later by Elliott Lieb and Walter Thirring.
Lenard and I consoled ourselves by remembering the words of my old teacher John
Littlewood at Cambridge: "First-rate mathematicians find bad proofs and then
second-rate mathematicians find good ones." But in this case, Lieb and Thirring
were also first-rate mathematicians.

Last Sunday Kitty Oppenheimer telephoned very distraught, saying she did not believe the doctors were telling her the truth and asking me whether I could find somebody who would. I thought at once of Trudy Szilard, our friend in La Jolla, whose husband Leo had a cancer of the bladder which was completely cured with radiation. She is a medical doctor, and having lived through this crisis with Leo, she knows everything there is to be known about it. I telephoned Trudy's sister in New York and was delighted to hear Trudy herself answer. She is now in New York and easily accessible. I gave her number to Kitty, and Oppenheimer tells me they had an hour's conversation which did Kitty enormous good. Yesterday Trudy came down and spent half the day with us here in Princeton. She was as sweet as ever with the children and talked about her various activities, China, Vietnam, and so forth. She is a splendid person, and I was happy to have been able to introduce her to the Oppenheimers at this moment. Oppenheimer continues active and does not look bad to the outward eye.

MARCH 24, 1966

I had a telephone call from London, a film magnate called Roger Caras asking me to come to his studio to help them with a science fiction film called *Encounter 2001.* Stanley Kubrick, who directed *Dr.*

Strangelove, is also doing this one. Caras will send a car to meet my flight from Paris and another to bring me to Winchester in the evening. I will stay with you a day or two longer to make up for it.

Since I was driven directly from the film studio to my family in Winchester, there is no letter describing the day that I spent with Stanley Kubrick watching him produce the film that became 2001, A Space Odyssey. *It is a puzzling film, totally different in style and subject from* Dr. Strangelove, *but equally memorable. I always considered film to be the most creative art form of the twentieth century, and Kubrick to be one of the great artists.*

MARCH 30, 1966

I am now finding out how lonely the Oppenheimers really are in spite of their huge number of "friends." I feel oddly more sad leaving them for two weeks than leaving Imme and the children. These are the last two weeks of Robert's radiation treatment, and in this time he must know whether it is life or death. I have been over three times to talk with Robert and Kitty. Kitty believes, perhaps rightly, that I can help Robert to keep alive by keeping alive his interest in physics. She feels desperately that he needs to be convinced that he is still needed in the community of physicists. On the other hand, I find that Robert is so physically tired from the radiation that my instinct is to hold his hand in silence rather than burden him with particles and equations. It is odd that I feel so personally responsible for him. I never had been close to him until now. I suppose it is partly the heredity that runs in our family that makes me want to save souls.

My mother's mother Eleanor Atkey was a famous faith healer, and my mother inherited some of her talent. Our daughter Mia, who is a Presbyterian minister, has it too.

Carl Kaysen spent a good two hours talking with me alone. I talked for an hour about the institute as it is, and he talked for an hour about the things he hopes to do with it. We understood each other very well. I shall enjoy having him for boss. He made one memorable remark: "I did not come here to operate a motel."

I had recently been interviewed by a journalist about the institute, and I said, "The institute is a motel with stipends," a remark that attracted some attention. Kaysen objected strongly to my description. Like the first director, Abraham Flexner, Kaysen wanted the institute to be active in public affairs. He did not want it to be merely a rest home for scholars. Kaysen's plan was to add a new school to the institute, a School of Social Studies, bringing together a group of people dedicated to understanding, and helping to solve, the problems of modern society. Like Flexner, he met furious opposition from the faculty but succeeded in establishing the new school which still flourishes today.

APRIL 22, 1966

In Germany we had a warm welcome, and I felt more than ever one of the family. This time I saw them in a crisis, and I saw them as they really are. On the first day I had a talk with *[Imme's brother]* Diether, and he explained what has been happening to Dr. Jung. Dr. Jung had recently come out of the hospital after his second operation. The first operation was an ordinary gallbladder removal. The gallbladder was found to be cancerous, so they did a second operation to look for metastases and take out lymph glands. They were able to take out everything that looked unhealthy. But Dr. Jung as a doctor knows that the chance of real success in such an operation is low. During our five days in Germany he was visibly growing stronger day by day. He was also having intermittent attacks of pain. I felt at first horrified to see him suffering, but in the end I felt strengthened by his calm bravery.

MAY 1, 1966

I have seen Oppenheimer several times since we returned, and each time he looks better and stronger. He is now recovering from the effects of the radiation. The cancer shrank to a size which is barely visible, and the doctors say they can remove it with a relatively minor operation. He is now working half-days in his office.

JULY 12, 1966

I had a phone call yesterday to say that Dr. Jung died. Imme had gone over there in a hurry with the three small children, so she did not come too late. It is a relief to know that this is over. Imme said she intended to ask you to go over there for the funeral, but it all happened so quickly that there was no time to organize anything. I went to see Oppenheimer yesterday and found him considerably better. Next week he will move out of his official residence and will be our next-door neighbour.

JULY 18, 1966

The Oppenheimers are now moved into the Yangs' house. I enjoy having them for neighbours. Kitty comes and gives me expert advice about my garden, which bushes to prune and which to eradicate. Robert is improving slowly. Yesterday we sat in the garden over drinks, and he was chatting for an hour and a half. Last night I took a bottle of champagne over to the house of my collaborator Andrew Lenard, and we had our little victory celebration which is also a good-bye, for he will be gone to Indiana before I return.

JANUARY 31, 1967, CANBERRA, AUSTRALIA

I came back from our three-day weekend in the snowy mountains. We saw a live wombat and a dead kangaroo. But for me it was the human fauna that were the most interesting. The Seymours, English immigrants, have been in Australia fifteen years without changing their characters or habits. They have one adored little boy, Martin, who is the center of their universe. But these good English parents, although they love and care for their Martin so intensely, are not able to express love in a way that a child can understand. He is like a guinea pig in a cage, carefully fed and admired but never handled. So when I came along and gave him a few bear hugs, the effect was electric. He responded passionately, and I became Uncle Freeman for the rest of the trip. He came running with me at every stopping place, and his parents remarked in a rather baffled tone of voice, "Oh, you are so good with him." I did not try to explain to them that their beloved Martin is emotionally starved.

FEBRUARY 16, 1967

We had two faculty meetings. Kaysen is pleased because I steered through the faculty the appointment of a new professor, a friend of mine from La Jolla [*Marshall Rosenbluth, a world leader in the field of plasma physics*], and this was done without any hostility from the side of the mathematicians. This was Kaysen's first new appointment, and he was waiting for trouble which did not come. After this good beginning, there is a chance that things will go smoothly for him. Oppenheimer is coming close to his end. He insisted on coming to this faculty meeting, but he can barely speak. We told him how glad we were that he came, but really it is a torture for everybody to watch him sit there speechless and suffering. His doctors have given him up, and we can only hope for a quick end. I had a long conversation with Kitty when I came home from Australia. She sounded close to collapse, and said she hardly sleeps anymore, but she was coherent and rational. The daughter Tony has given up her university studies and taken a job in Princeton so that she is constantly available. Imme remarked that Tony had better be careful or she will find herself nursing her mother for the next twenty years. After saying good-bye to his New York doctors, Oppenheimer has put himself in the hands of our Dr. Blumenthal in Princeton, and Kitty says that Blumenthal has been splendid. Both medically and as a human being, Blumenthal would be a good man to have by when one is dying.

Robert Oppenheimer died in his sleep in the evening of February 18. Kaysen spent the following days organizing a memorial ceremony and taking care of a large number of distinguished guests who came to pay their respects to Robert. Among those who came was General Leslie Groves, the leader of the wartime Manhattan Project, who had personally picked Oppenheimer to direct the bomb laboratory at Los Alamos. Groves and Oppenheimer were as different as two people could be, but they worked harmoniously together to get the job done. After the ceremonies were over, with moving reminiscences from George Kennan and Hans Bethe and other friends and colleagues of Oppenheimer, Kaysen remarked to me, "Well, if they fire me from the institute, I can always make a living as a funeral director."

APRIL 23, 1967

On Friday came the news of Stalin's daughter. [*Svetlana Alliluyeva, the daughter of Stalin, had been allowed to leave the Soviet Union and was living a restless life as an exile, first in India and later in Switzerland. Various people in Princeton with an interest in Soviet history were anxious to talk to her.*] Edward Greenbaum is a friend of ours and a trustee of the institute. He happens to live next door to Helen Dukas, and so Helen knew when he went off on a secret mission to Switzerland a week ago. Helen immediately guessed that he would be bringing Stalin's daughter back with him. Nobody believed her until it was in the newspapers on Saturday. Now she is enjoying her triumph. We were glad that Kennan played a courageous role in this affair. He has been as a historian intensely interested in Stalin and took the initiative in going to see Svetlana in Switzerland and persuading her to come here.

JUNE 10, 1967

This week has been spent in a state of wild excitement over the Middle East war. Princeton being a heavily Jewish town, many of our friends have brothers and sisters and aunts and uncles "over there." I was never very friendly to Israel. [*I had lived in England through the years when British soldiers were stationed in Palestine, trying to keep the peace between Arab and Jewish populations, and our soldiers had been repeatedly attacked by Jewish guerrillas.*] But this war leaves me profoundly grateful to these people for managing everything so well without our help. I never believed it was possible in this century to fight and win a war in four days. It seems there has been nothing like it since Agincourt. On the first day of the war Dorothy had a lot of her pictures exhibited at an art show at the Princeton Synagogue. In the same school there is a little Israeli girl called Tamar, and we met Tamar with her mother at the show. Tamar is also very gifted. Her mother was saying good-bye to everybody because she was flying home to Israel on Tuesday. I thought, how irresponsible to take a child like that into the middle of the maelstrom. But the mother was completely confident. Another lady offered to keep Tamar in Princeton for a few weeks, and the mother refused without hesitation. I have to admit now she was right.

Our sixth child, Rebecca, was born on July 2. There is no letter describing the birth. The transatlantic telephone had become cheap and convenient enough, so that important events would be reported by telephone. When we first told the girls that a new member of the family would be arriving, Dorothy said, "No, that is wrong. If a new baby was coming, it would have come last year. We had babies in 1959, 1961, and 1963, so the next one would have come in 1965." We had to explain to her that babies are sometimes allowed to break the rules. Rebecca was an easy baby and grew up with a startling resemblance to childhood pictures of my mother. The same telephone calls that told my mother of Rebecca's birth brought to us the sad news, that my mother was going to hospital for an operation with a suspected colon cancer. Like Dr. Jung, she needed two operations. Unlike Dr. Jung, she survived them both and lived for seven more years. The following letter was written in response to the alarming news from my mother.

JULY 27, 1967

Dear Mamma, I was thinking, what a bright and charming little girl you must have been in the days when you were given young larks and partridges to take care of, and now there is not one person left alive in the world who remembers you as you were then. All the time I am enjoying the wise talk and tender love of the three little girls, Dorothy's shrewdness, Emily's sweetness, and Miriam's mischievousness, and I am thinking how evanescent this all is. You are perhaps coming near to the end of the road, and we are at the beginning of it, but it is the same road for all of us.

In these days I think of the years when I was close to you and spending many days walking and talking with you, the years we lived in London until I went to America, from 1937 to 1947. I was lucky to have you then to see me through the years of Sturm und Drang, to broaden my mind and share with me your rich knowledge of people. I remember reading aloud with you *Sons and Lovers* by Lawrence, knowing that you and I were a little like Lawrence and his mother, and that this perfect intellectual companionship which we had together could not last forever. I am grateful for those years, and for the many weeks of renewed companionship that we have had since. I hope this letter does not sound too solemn. I expect to come and see you many times still in the future. I write you these thoughts, just in case you should slip away.

July 30, 1967

Did you notice that you and I both used the phrase "if you should slip away" in the letters which we wrote to each other on July 27? I am quite willing to believe that some kind of telepathy may happen at moments of intense feeling such as this. I would like to believe it, because it would be consistent with the idea that some kind of world-soul may exist, an idea that has appealed to both of us. I am a total agnostic, but the laws of physics and chemistry do not exclude a world-soul any more than they exclude our individual beings.

In 1967 Esther and George were still happy to baby-sit and fool around with their little sisters, but they were moving rapidly into a different world. They were teenagers and needed to take charge of their own lives. They were growing up and metamorphizing like caterpillars. You do not know, when the caterpillar becomes a pupa, whether it will emerge as a beetle or a butterfly. Esther applied for early admission at Radcliffe College and got a letter saying she was admitted but strongly advised to wait a year before enrolling. She fought back and enrolled at age sixteen. George at age fourteen became silent and withdrawn.

SEPTEMBER 16, 1967

Today I am driving Esther with ten pieces of luggage to Boston. Esther had her Youth-of-the-Month award. More exciting to me was a certificate from the New Jersey authorities certifying her to be the best high school Russian student of the year for the whole of New Jersey. George is hardly visible. He spends much of his time at the Catacomb, a club for teenagers in the cellar of the Presbyterian church. When at home he mostly plays very quiet sithar music. His bedroom is furnished in the style of a Hindu temple, and with the sithar music and constantly burning incense, the Indian tapestries and paintings covering the walls, the effect on the nerves is very soothing. I think he is beginning to find himself in this pseudo-oriental milieu.

NOVEMBER 17, 1967

It is your golden wedding day, and I cannot let it pass without writing. You can have some satisfaction in looking back, to see how much better the world is in so many ways than it was in 1917. That was a low point in human hisory. You have much to be proud of. Two satisfactory children, forty-six years of sustaining Papa through the ups and downs of a productive life. It is good that you were able to see him through to the end. He would have been much more lost without you than you are without him. You have been the rock on which he built his life.

My parents were married on November 17, 1917, when World War I had destroyed a whole generation of young people, the Germans had won huge victories on the Eastern Front, and the end of the war was not in sight. My father was lucky to have been sent home from the Western Front as a mental casualty, suffering from the disease that was then called shell-shock and is now called Post-traumatic Stress Disorder. My mother nursed him back to health.

Last night I was in Ithaca at the banquet to celebrate Hans Bethe's Nobel Prize. Ithaca was beautiful, already under snow. Hans was entirely himself, as imperturbable as ever. Rose Bethe said she is glad the prize came to them so late. She said if they had had the prize ten or twenty years ago, many of their friends would have been slightly jealous or doubtful about it. Now everybody considers it overdue and is genuinely happy about it. She is right. It was an outstandingly warm and happy occasion.

DECEMBER 26, 1967

Esther turned up here for two days. I learned a few things about her life at Radcliffe. The main thing is, she has been elected to the *Crimson* board. As usual, she got what she wanted. The board consists of thirty people of whom only two are first-year students, from Harvard and Radcliffe together. So she had stiff competition to get in. She seems to be firmly set on a career in some kind of journalism. For this the *Crimson* is an excellent starter. However, I told her in the best fatherly tone that if all she wants from Radcliffe is on-the-job training in journalism, she should take a job on a real newspaper right away and save me about $15,000. She agreed she will now let the *Crimson* slide a bit and concentrate more on her formal education.

JANUARY 25, 1968

Imme and I had a hilarious evening at the Kaysens. We were invited to supper with Jim Watson (the DNA man) and other people. It was an informal affair, a different atmosphere from such parties in Oppenheimer's time. The guests were all Harvard people, and there was much talk about Cambridge, Mass. Kaysen started out by saying,

"Well, I hear that your Esther has taken Cambridge by storm." Being elected to the *Crimson* has really impressed people there. Kaysen said, if you want to get a good education at Harvard, the thing to do is to go onto the *Crimson*. What I find impressive is that Esther understood so fast that this was the thing to do, quite apart from her performance in actually doing it.

Meanwhile we are watching George move in a very different direction. He has as quick an eye for the road to perdition as Esther has for the road to success. He chooses his friends among the hard-core rebels whose main amusement is taking illegal drugs. Possibly he gained entry into this circle, as difficult in its way as breaking into the *Crimson*, by bringing some supplies from San Francisco. The hardcore drug circle has a rigid code of honour, the basic rule being *omertà*, silence, not to talk even under the severest pressure. If they accept George, he must have in some way proved his worth. The other night I had a real father-to-son talk with George, the first in several years. We talked for more than two hours. I told George as emphatically as I could of the many young people I have known who went to pieces as a result of drugs. He replied that he knows many more drug-takers than I do and that they all lead sane and satisfactory lives. That was round one. We continued arguing passionately but politely on both sides. In the end I told him that I will do everything in my power to stop him handling drugs, including bringing in the police if I get any hard evidence. I said I would rather see him in gaol than in a mental institution. He took this all calmly, said he respects my frankness and hopes we can continue to respect each other on both sides. One thing at least is clear as a result of this conversation. George is as formidable in his way as Esther is in hers.

February 15, 1968

I am still hopeful that George will come through all this and make something good of his life. I enclose a couple of poems which he wrote recently, probably in a marijuana trance. They are technically imperfect, certainly not on the level of Coleridge's "Kubla Khan," but I enjoy the dreamlike flow of the words and images.

> *And the red knight he tottered from his silken velvet throne*
> *Because the fair one he had wanted was to be no more his own.*
> *And now that he was gone the place was opened to despair*
> *When up upon the hill there cried a youth with long black hair,*
> *And often did he scream, and often he would yell,*
> *That because of the young and fair one, the world would rest in Hell.*
>
> *And often did they speak of silver silk and gold*
> *To come in times of peace and happiness untold.*
> *And Gods would win their fortunes and Kings let down their hair*
> *Till times would come again when evil was laid bare.*
> *And the ground would bring forth flowers of colors yet unknown*
> *To bathe our mind and body in thoughts our very own.*
>
> —GEORGE DYSON, *January 1968*

Things went much faster than we expected. On Tuesday two big policemen came to the house and asked if they could search George's room. Then they picked him up directly from school. He has been for three days in the youth house in Trenton, and we are due to get him back tomorrow. In the evening one of the two policemen came again to talk to Imme and me. He seemed to us a very sound and understanding person. He is youth officer for Princeton and handles all the juvenile delinquents here. He said the first thing they did to George at the youth house was to cut his hair. This will cause George tremendous grief. As you can see from his poems, the long hair is important to him. As the policeman said, they do not want to maltreat the boys at the youth house, and yet they want to make sure that the boys really try to keep out of it. This symbolic indignity of the haircut is an essential part of the treatment.

I heard good news of George from a lady who was with him on

a camping trip in the High Sierras. She said George was the leading spirit in a group of teenagers who were at the camp and impressed everybody with his flamboyant clothes and behaviour.

Last weekend Esther turned up here, on her way from a reporting trip to Haverford College in Pennsylvania. I took Esther down to the Youth House to have an hour with George, who was still incarcerated. The youth house is a little modern building in Trenton. It is boring rather than punitive. George said he was not badly treated there. The following day he was released and came home looking greatly improved with his haircut. On Tuesday his case will come to the juvenile court, and the court will no doubt put him on probation. I am glad they caught him so early.

MARCH 24, 1968

Last week Imme and I had to appear at the courthouse for a social investigation, which meant that each of us plus George had a half-hour interview with a benevolent but not very bright social worker. Next week there will be a psychological examination. After that finally will be the second hearing by the court, at which the judge will announce his decision. The whole procedure is made inconvenient and time-consuming for the parents, because in the majority of cases the parents are negligent or uncooperative, and this is a good way to convince them that it is worth their while to keep their kids out of trouble.

APRIL 5, 1968

To my mind the shooting of King was a far worse thing than the shooting of Kennedy. I heard King speak in Berkeley about fifteen years ago, before he became famous, and I always had a great belief in him. He was far and away the greatest and most far-sighted of the Negro leaders. I do not blame the negroes at all for rioting now. If I were black, I would be out in the streets with them. Esther talks of going to Europe this summer at her own expense. George is not really changed. Still surly and insolent by turns, but at least he is now working at his schoolwork and getting some A and B grades. The psychologist pronounced him only normally maladjusted.

APRIL 21, 1968

I ran into a problem which you may be able to elucidate. I wanted to refer to the scene at the end of *Faust* where Faust is talking about organizing a group of people to drain a swamp. I had in mind a vivid picture of this scene, Faust with a shovel in his hand working at the dike, finally taken out of himself by the comradeship of a common effort. When I looked for this scene in the play, I found it is not there. What is there is a philosophical discourse rather than a dramatic picture. Looking back into the past, I wonder where this beautiful and spurious Faust-scene came from, and I believe it must have come from Mamma. Do you have any recollection of having told it to me? And if you did, was your improvement of the original version conscious or unconscious? It is a good illustration of the limited value of education, when the only piece of Goethe that made a deep impression on me turns out to be by you and not by Goethe.

The institute is lively, there are ten physicists here and I am working hard at some good problems. Today I discovered a little theorem which gave me some intense moments of pleasure. It is beautiful and fell into my hand like a jewel from the sky.

This theorem was published in a little paper with the title "A New Symmetry of Partitions" in a mathematics journal (1969). Through all my years in America, I took occasional short holidays from physics and returned to my first love, the theory of numbers.

JUNE 30, 1968

I was invited to a meeting of the Defense Science Board, a high-level body of which I am not a member, to discuss nuclear policy. I do not have the illusion that my words at a meeting like this can change anything much. But at least I had something definite to say, and I said it. In general terms, what I had to say was, "Remember 1914." The professional soldiers spend so many years working out their elaborate war plans that they forget the essential difference between plans and reality. Perhaps I helped to bring them a little closer to reality. It is an odd thing to go to a meeting like this where one tries to influence decisions that may mean life or death to whole countries, particularly in Western Europe. If I have had any good effect, I will never know it. The meeting was thoroughly enjoyable on a personal level, everybody calling one another Fred, Freeman, Sid, etc., and half the time joking. It might have been a lively college debating society. But for me at least, after my experiences at Bomber Command twenty-five years ago, it was not just an intellectual exercise.

In the spring of 1969 I spent a term teaching at the University of California at Santa Barbara. I slept in a guest room at the faculty club. That spring was a time of great turmoil at many universities, including Berkeley, Santa Barbara, and Harvard, with rebellious students going on strike and occupying academic buildings to protest the Vietnam War. The violence was worst at Berkeley, where police invaded the campus and many students were arrested. At Santa Barbara and Harvard, protests were mostly nonviolent.

APRIL 12, 1969, FACULTY CLUB, UNIVERSITY OF CALIFORNIA, SANTA BARBARA

I was wakened at six-thirty in the morning by a tremendous crash, followed by some shouts of "help." I thought somebody must have driven a car into this building at seventy miles an hour. I discovered then that I am after all a coward. Instead of running out immediately to the rescue, I took about a minute to pull myself together, to face whatever had to be faced. In that minute I was somehow paralyzed. And in that minute a man burned to death. I will not forgive myself for this.

When I did run out, I found it was not a car crash but a bomb explosion. Two men were already helping the injured one, and they had quickly carried him into an ornamental pool which put out his burning clothes. He was sitting there in the pool, and he did not look too bad. I could see only that his legs were burned and one hand injured. I then telephoned for an ambulance but somebody had already called. In a few minutes the ambulance men came and put him on a stretcher and took him away to the hospital. The fire in the faculty club was easily put out. At that point I assumed that we had been lucky. Only later I heard that the man was burned so extensively that he is not likely to live. And I cannot escape some responsibility for this. The man was the caretaker of the building, and he found a large package lying in front of the door. It was booby-trapped to explode when he opened it. It was made with a stick of explosive in a container of petrol so that he was well showered with burning petrol. It seems unlikely that we shall find out who placed the bomb. Undoubtedly the radical students will be blamed for it. They have organized a "free university" in the student center next door, including courses in guerrilla warfare and the manufacture of homemade weapons. They have also been protesting the existence of the faculty club as an infringement of their rights.

What is one to make of all this? I look down now from my window, and there is the little pool only ten feet below, where the man was sitting yesterday. Around the edge the blood and ashes are still there. He did not bleed much. If I had been a little quicker, I could have dragged him into the pool in a few seconds. But I did not even look out of the window until it was too late.

April 20, 1969,
Faculty Club, UC Santa Barbara

It seems a long time ago that Dover Sharp was sitting burned in the little wading pool. He died on Sunday a week ago. I was called in by the police investigating the murder, but I was not able to tell them anything useful. Today the children were running and splashing in the little pool just as if nothing had happened. I remember one of William Blake's proverbs that I used to be fond of when I was young: "Drive your cart and your plow over the bones of the dead." They certainly believe in that here.

It is interesting to me that you feel more confident about George than about Esther. I would say that this is because you do not know George. I admire the courage and independence of each of them. But George is cursed with his mother's temperament while Esther has mine. He will always do everything the hard way. I hope for the best, but I will be surprised if he comes through a year on his own without serious trouble. I do not know whether George got accepted at Harvard. Harvard itself is in such a mess just now that we may not know for some time. I had a long and full letter from Esther describing her role in the events of this week. At the mass meeting in the football stadium, Esther went to the microphone in front of ten thousand students and made a speech arguing for peace and moderation. She also rebelled against the majority on the *Crimson* editorial board who were in favour of coercion. I think she is all right. At least she is getting an education out of this, if not in the way I had intended. And she is clearly on the side of sanity.

May 10, 1969, UC Santa Barbara

I spent the last few days writing a testimony about missile defense for the Senate Committee on Armed Services. I am in favor of defense but almost all my friends are against it. The crucial Senate vote on the question will come in about a month from now. I wrote a dry and technical testimony without much hope that it will convince anybody. But then I received a copy of the testimony of Wolfgang Panofsky, the chief scientific witness against Missile Defense. He is the son of our old neighbour Panofsky who died two years ago. Panofsky made a good statement, solidly backed up with facts and figures. But at the end of his statement Panofsky argued that we do not need a missile defense against China because we can always destroy the Chinese missiles with a preemptive strike if the necessity arises. When I saw this I said, "Now the Lord hath delivered him into my hands," like Thomas Huxley in his famous debate with Bishop Wilberforce about the descent of man. I put into my statement the following paragraph:

> I am amazed at the cheerfulness with which some of my scientific colleagues, arguing against the deployment of missile defense, speak of our ability to launch a preemptive nuclear strike against China. Anybody who considers a preemptive strike to be preferable to missile defense has not understood what the Cuban missile crisis was all about. The whole point of the Cuban missile crisis was that President Kennedy succeeded, with great wisdom and some luck, in finding a way to avoid a preemptive strike. Precisely to enable some future president to play it cool like Kennedy, to resist pressures to make a preemptive strike in a moment of desperate crisis, this is to my mind the main purpose of missile defense. It is not so much to save our skins as to save our souls.

This paragraph came to me like a flash of lightning, and I think it will get through to the senators. It is true, and it is in language they cannot fail to understand. Especially it is designed to speak to Senator Edward Kennedy who is the leader of the opposition to missile defense. If you read in the newspaper that Senator Kennedy changed his mind, you will know why.

Senator Kennedy did not change his mind, but the United States continued to develop and deploy missile defense, with my support. I support missile defense as a long-range strategy to replace offensive nuclear forces and possible preemptive strikes. Instead, we have deployed missile defense in a halfhearted way, as an addition to offensive forces and possible preemptive strikes. I still have hopes that we will one day shift to a genuinely defensive strategy, getting rid of offensive forces and using missile defense as the basis of a more stable world.

May 20, 1969

George has decided to undertake an extremely rugged walking tour alone. He will walk down the Continental Divide over the highest part of the Rocky Mountain country. It will take him forty days, and I suppose he is aware of the biblical precedent. Imme and I talked with him at length about the details of his plan. He has tried to foresee all the difficulties. He will nowhere be further than three days walk from some kind of road. He has ordered seventy large-scale maps which he will carry along with him, and also seven smaller-scale maps which he will leave with us with his intended route marked on them. So if he is missing, we will know where to send a rescue party. He is collecting an elaborate medical kit so he can deal with anything from diarrhea to snakebite. He has also a fat book with pictures of all the edible and poisonous plants of the Rocky Mountains, which he will carry so he can live on the country. Imme and I agreed that we should let him go. The risks are real. But it is obviously necessary to George's spirit to do something extraordinary. We are impressed with the professional way he is making his preparations. When I said good-bye to him this morning he looked grateful and happy.

When George was preparing his trip, I asked him whether it would not be safer to take a friend with him. With two people, if one was sick or injured, the other could walk down to find help. George replied, "I would be delighted to take a friend along if there were anybody I could stand to live with for six weeks."

July 6, 1969

George met ferocious and continued snowstorms, quite unusual at this time of year, and the high country was so deep in soft snow that it was completely impassible. He struggled for a week and then retreated to Chicago. I am glad that he had the sense to retreat when things became impossible. He will start again further south in a few days from now and hopes to do the second half of his originally planned route. I had two good letters from him from up in the mountains. He may have learned something from this fiasco.

AUGUST 9, 1969

We spent two days glued to the TV watching the moon trip. For me the moon program came as a surprise. I had expected they would do their best to make it into a public spectacle, but I had no idea this would be possible on the very first trip. I had known what the astronauts would be doing, but I had not imagined that they could have a TV camera set up in the right place so that one could see all this as it was happening. I was astonished at the way the astronauts talked. Normally they say very little when they are working, but for the whole two and a half hours on the moon they kept up a continuous stream of conversation, full of information and clearly understandable. The operation was obviously planned as a theatrical production with the whole world for audience, but I had not imagined it could be done so effectively. From a scientific point of view I am much more excited by the Mars pictures which came back this week. Both the Mars flights were brilliantly successful and have given about a hundred times more information about Mars than we had before.

George came to see me in Boulder the day before I left. He spent altogether only two weeks on the high mountains instead of the six he had planned. The first time he went up, there was too much snow to move. The second time he found the going very easy, and he covered twice the distance he had expected each day. Now he is at the Sierra Club camp and earning his keep there.

SEPTEMBER 11, 1969

I quote a postcard which arrived for George yesterday: "Dear George, wherever you are, thank you for your great work in the San Juans. Your initiative and effort and attitude made the camp operate very smoothly. Whenever you get your check, it will include a bonus in appreciation of your help. Hope you'll be available for the Wind Rivers next year."

The postcard came from the manager of the Sierra Club camp where George was a counselor. At this camp he got to know Barbara Brower, then a teenager, the daughter of the Sierra Club director David Brower. As a result, he was invited to visit the Brower home. He was informally adopted into the Brower family and became a friend of David Brower, Anne Brower, and their sons Kenneth, Robert, and John, who became for him the brothers his own family lacked.

NOVEMBER 30, 1969

A thing happened this week which makes me particularly happy, the announcement of Nixon that he will renounce and destroy all biological weapons. This was largely brought about by a friend of mine, Matthew Meselson, who shared an office with me at the Disarmament Agency in Washington six years ago. Meselson is a Harvard biologist. He spent a great part of his time for the last six years quietly campaigning against biological weapons. Finally his efforts came to fruition this week. I wish I had been equally effective.

Nixon was able to move decisively to get rid of biological weapons because he made the move unilaterally. He did not need to negotiate details of the move either with the Soviet Union or with the U.S. Senate. Twenty years later George Bush senior made a similar unilateral move getting rid of half of our nuclear weapons. Bush removed all the nuclear weapons from the U.S. Army and from the surface navy, leaving only the nuclear weapons belonging to the air force and the submarine navy. These two unilateral moves were the most substantial acts of disarmament in our history. Unfortunately our academic experts are always talking about disarmament reached by international negotiation rather than unilaterally. In my opinion, major and important acts of disarmament will always be easier to make unilaterally. It was also an advantage that both Nixon and Bush were right-wing Republicans. For a Democrat, such a unilateral move would be much more difficult.

Esther telephoned that she is hoping to have a summer job working for Senator [George] McGovern, one of our most intelligent and peace-loving senators. It would be a great experience for her if she is accepted. A brief letter from George announcing that he is happily settled at Berkeley and is living on board a twenty-two-foot sloop in the San Francisco Bay. It sounds like an idyllic existence. He is intending to buy the boat and eventually to sail her. She was built in England nine years ago and came to San Francisco under her own sail. So one day you should expect him to come sailing up Southampton Water to pay you a visit. I hope he will learn how to sail before trying any big voyages.

Soon after I received this letter from George, he disappeared from Berkeley. His sister Katrin was then living in Vancouver. With some help from Katrin, he went to join her in Vancouver. I told him that I wholeheartedly approved of his move to Canada. At that time Katrin had another friend, Jason Halm, who had been drafted into the U.S. Army and was already on his way to Vietnam. Jason was in the back of an army car driving through San Francisco to meet the boat taking him to Vietnam, and at the last possible moment he quietly dropped out of the car. With some help from Katrin he also arrived in Vancouver. Soon after that Jason and Katrin were married. They invited George to attend their wedding, and during his visit he answered a newspaper advertisement for a job on a boat, and stayed. George and Jason were welcomed generously by the Canadian government and later became Canadian citizens. George moved out of the Halm household and built himself a comfortable home a hundred feet up in a tall Douglas fir tree overlooking the ocean a few miles north of Vancouver. A few years later, Kenneth Brower published The Starship and the Canoe *(1978), which described George's life in Canada.*

FEBRUARY 28, 1970

I was taking care of Stephen Hawking, a young English astrophysicist who came here for a six-day visit. I had never got to know him till this week. Stephen is a brilliant young man who is now dying in the advanced stages of a paralytic nerve disease. He got the disease when he was twenty-one and he is now twenty-eight, so his whole professional life has been lived under sentence of death. In the last few years he has produced a succession of brilliant papers on general relativity. In conversation he has one of the quickest and most penetrating minds I have come across. He is confined to a wheelchair, can barely hold his head upright, and his speech is hard to understand. These days while Stephen was here, I was in a state of acute depression thinking about him, except for the hours when I was actually with him. As soon as you are with him, you cannot feel miserable, he radiates such a feeling of strength and good humour. I was running after him to escape from my misery. After spending these days with him, I am not surprised that he found a girl who would marry him. They have a three-year-old son about whom Stephen talks with great pride. Stephen would only laugh if you told him this, but I think he must be some kind of saint.

By some miracle that the doctors do not understand, Stephen Hawking is still alive and still intellectually active at age seventy-four. The last time I saw him was at a meeting in New York in April 2016, discussing future space missions to be funded by the Russian oligarch Yuri Milner. Totally paralyzed and able to speak only through a computer, he still travels around the world, writes best-selling books, and enjoys being a public celebrity. The public has good taste in its choice of heroes. The public responds to Hawking as it did to Einstein, knowing that they are great human beings as well as great scientists.

20

ADVENTURES OF
A PSYCHIATRIC NURSE

WHEN I WAS appointed a professor at the Institute for Advanced Study, I used to say that my real job was to be a psychiatric nurse, giving consolation and comfort to the young visiting members when they suffered from loneliness or depression. The visiting members were in a highly stressful situation, facing a year or two of complete freedom, with the expectation that they should do something brilliant. If they failed to perform, given this unique opportunity, there was a real danger of psychological collapse. In my time as a professor I lost three young people whom I had invited as members, one by suicide and two who ended up in mental institutions. I do not know how many I saved. I only know that the institute is a dangerous place for young people, and as a professor, I bore a heavy responsibility for their mental health. The letters are as usual arranged chronologically, beginning with family affairs and then telling stories of psychological disasters.

One of the group of professors arriving at the institute in 1935 was Hetty Goldman, a famous archaeologist who spent many years excavating the ancient city of Tarsus in Turkey. When I came to Princeton, Goldman was retired, but the institute maintained an active group of archaeologists led by Homer Thompson. Thompson organized archaeological lectures for the general public. The lectures were well advertised and well attended. My hometown Winchester, where my parents lived, sixty-one miles southwest of London, was also a famous archaeological site. Our house was built over a Jewish cemetery with graves of wealthy Jews who flourished in the city in the twelfth century A.D. Anywhere in the neighborhood, if you dig down a few feet, you will find historic relics. Winchester began as a Celtic city around 300 B.C., was enlarged by the Romans and further enlarged by the Saxons. It was the capital of the Saxon kingdom of Wessex where King Alfred reigned in the ninth century. After the Norman conquest, Winchester remained an important center of the royal administration, with an enormous cathedral and an enormous palace for the presiding bishop. The college where my father taught and I studied still occupies the solid stone buildings built when it was founded in 1382.

MARCH 17, 1971

On Monday there was an evening archaeology lecture at the institute given by Martin Biddle, who has directed the digging in Winchester for the last ten years. He gave a brilliant talk. The professional archaeologists were enthusiastic about the technical quality of his work. Our friend Homer Thompson, who has directed the American diggings in Athens for the last twenty-five years, said that the Winchester project is of comparable scope and quality. For Imme and me, it was exciting to see the beautifully clear slides of the places we know so well. There was a good one of the Barracks with [*my parents' home*] 1 St. James Terrace clearly visible behind it. Unfortunately we could not see your faces at the windows. The Barracks are built on the site of a huge double-walled Norman castle which is not yet excavated. There was also a fine view of Oram's Arbor which has underneath it the main western entrance to the pre-Roman city of 100 B.C. From the point of view of the historians, the most exciting discovery is a number of graves and personal belongings showing clearly that there was a considerable Saxon influence, and probably a Saxon population, in the city before the Romans left. It seems the Saxons did not originally come as invaders but were invited over by the Romans to help defend the city. Then, when the Romans were gone, the Saxons brought over more of their friends and relations. It all makes sense.

In school we had been taught that the Romans left a Celtic population in Brit-
ain that was afterwards overrun by hordes of invading Saxons. This version of
history turned out to be wrong.

MAY 22, 1971

The big event this month was the talent show at Johnson Park School, organized by the new music teacher. Four days before the show he sent us a message saying that Emily must have a white dress, so Imme started furiously working at her sewing machine and got a very pretty dress finished in time. The show began with a science class, the teacher Mr. Twiddlewiddle being splendidly acted by one of the fifth-grade boys. He was telling the class in very dry language about the solar system. Then a bell rang, and Mr T. was called to the principal's office. He said to the class, "You go on working at your science projects while I am away." One of the boys brings in a space machine that he has built so that the class can tour around the solar system and investigate whatever forms of life exist there. They find on each planet a form of life that remarkably resembles some form of life on earth. On Mars there is a rock-and-roll band, on Venus a group of girls singing, "I got to wash that man out of my hair," on Mercury a group of baton twirlers, and so on, each item done fast and well drilled. When it came to the moon, a silvery light filled the stage, and there was Emily alone with her white dress and her violin. She played "Moon River," an old popular tune which I had worked hard at practicing with her. She looked lovely and also played well. The hall was packed and everybody praised her. After the tour was finished, the class came back to the classroom, and Mr. Twiddlewiddle returned. The class all began excitedly telling Mr. T. about the forms of life they had discovered. "Sorry, we have no time for that," said Mr. T., and went on with his boring lecture.

JUNE 18, 1971

Helen Dukas had a delightful letter from Svetlana. Helen had said, if Svetlana will come to visit Princeton, Helen will be glad to baby-sit for her, having so much good experience as a baby-sitter with us. Svetlana replied that the one thing she does not need now is a baby-sitter. With her two babies in Russia, she always had so many nurses and maids that she hardly saw the children except for official occasions. With this last baby she is determined to do everything herself. She nurses it, has it with her twenty-four hours a day, and is blissfully happy. She said she is also glad to see the baby looks more like her father than her grandfather.

Stalin's daughter Svetlana had stayed for a while in Princeton when she arrived in America in 1967. She married the American architect Wesley Peters and gave birth to her daughter Olga in 1971.

August 21, 1971

You ask what I think about the moon. I was glued to the television all the time the men were outside. This expedition was quite different from the earlier ones because this was the first time the astronauts were seriously doing science. They had absorbed a great deal of knowledge and were eager to understand the geology of the place rather than just collect trinkets. For the first time I had the feeling science was the honest purpose of their trip and not just window dressing. This place was far more beautiful and exciting to look at than the flat places the other astronauts had visited. I am sure they did so well this time because they know there will only be two more landings. It is a race against oblivion, to collect all the information they possibly can before the axe falls.

The mission that I was watching was Apollo 15, the first that carried a roving vehicle so that the astronauts could travel and explore over a considerable distance. The landing site was the Hadley Rille, a canyon with spectacular views of the surrounding mountains. The rover carried television cameras so that the worldwide audience could see the territory that the astronauts were exploring. For the first

*time we saw the lunar highlands, the unearthly landscape of mountains and val-
leys that cover most of the Moon.*

NOVEMBER 13, 1971

The week has been full of Einstein. I have been reading the new
Life of Einstein [1972] by Ronald Clark, an Englishman who was here
last year. I think Clark's book is good, but Helen Dukas is deeply
angry at Clark because of what she considers betrayal of her confidence. Clark printed a number of stories about Einstein which Helen
had vetoed. The book is already a best-seller here, but Helen and
Otto Nathan, the trustees of the Einstein estate, have refused permission for all English and European publication. Their lawyers and
Clark's lawyers are now fighting it out.

Meanwhile Banesh Hoffmann, an old friend of Helen, is at the end
of another biography of Einstein [1972]. Yesterday I spent the whole
day with Hoffmann going over it. Hoffmann will print nothing that
Helen does not agree to, and in return he gets a magnificent supply of
illustrations and documents taken from Helen's collection. It is now a
race to see whether Hoffmann or Clark gets there first with the European public. The two books hardly overlap, and there is no reason
why they should not both be successful. Clark's book is fat, crowded
with colour and detail, and gives the first honest account of Einstein's
many squabbles. He is on the whole fair to Einstein, but not every one
of the stories he prints on the basis of other people's memories need be
totally correct. Hoffmann has two advantages. He worked with Einstein and knew him well, and he understands the physics. Hoffmann's
book gives a personal view of Einstein as a working scientist, something that is outside Clark's compass. Also his book is about a quarter
of the size of Clark's, which I consider an advantage.

Into this tense and acrimonious situation walked a third character,
an Indian friend of mine called Jagdish Mehra. Mehra is a historian
of science who has one quality that both Clark and Hoffmann lack, a

gift for languages. He has talked at length with European relatives of Einstein, some of whom talk only German and some only French. As a result, he discovered last year in an attic in Belgium the manuscript of Einstein's first written work, an essay on electromagnetism which Einstein sent to an engineer uncle when he was sixteen years old. Nobody else had known that this existed. It was written in a difficult Gothic script which Mehra is able to read. Mehra was excited with his discovery. But from that moment his troubles began. He behaved carefully and correctly. He first copied the paper, sent the copy to Otto Nathan, and asked for permission to publish it in a scholarly journal. Nathan replied giving the permission. Then Mehra translated it into English and sent it with a commentary to the *American Journal of Physics*. There he ran foul of another historian of science called Gerald Holton. Holton advised the editor of the journal that Mehra's paper was inadequate and ought not to be published. Holton also claimed that the translation was faulty. So the paper was rejected. At this point, Mehra decided to send the thing to *Physikalische Blätter*, the German magazine, and so avoided the problem of translation. *Physikalische Blätter* accepted it gladly. Mehra was happy to have the thing finally settled when he suddenly got a brief note from Nathan saying, "Permission to publish withdrawn" without any explanation. Mehra sent this note to the editors of *PB*, but they decided it was too late to change their minds, and the Einstein paper appeared in September. Helen and Nathan regarded this as a direct defiance of their authority, and they have excommunicated Mehra from any further contact with any Einstein papers which they control.

Why did this happen? I think I know the true explanation. In Hoffmann's book, which Mehra never saw, there is an episode in which Einstein said, "After the massacre of the Jews in Germany I will never allow my writings to be published in Germany again." Such a statement would be for Nathan a command which he is pledged to carry out to the letter. Poor Mehra innocently committed the supreme sacrilege of publishing Einstein's first paper in Germany. Now I am trying to straighten out this mess. At least I am happy about one thing, that I am not a historian. These historians are so jealous of each other, it is unbelievable. How Einstein would laugh if he could see them quarreling over his relics.

DECEMBER 9, 1971

The Einstein story had a surprising sequel. Imme and I were at supper at the Kaysens last week. I happened to mention that Mehra had been staying with us. Kaysen immediately looked startled. He said, "Do you mean the Mehra from Texas? Just a month ago I almost put that man in gaol," and proceeded to tell us the details. Mehra came to Princeton in October and spent some time talking with Helen Dukas in the strong room where the Einstein papers are kept. Helen let Mehra look at the papers under her watchful eye. A few days later she came to Kaysen in terrible distress to tell him that one of the most precious manuscripts from the year 1914 was missing. She had seen the manuscript in its place only a week before. Kaysen went to a lawyer and got him to write a fierce letter to Mehra accusing him of the theft and threatening prosecution. Mehra wrote back nonchalantly denying any knowledge of the affair and protesting his complete innocence. A few days later Helen received a postal package containing the missing document, posted from Newark Airport. After that Kaysen decided he could not prosecute and wrote a personal letter to Mehra telling him that he should never again show his face at the institute. Kaysen put a detective on the job of tracing the origin of the package and determined that it had been brought from Texas to Newark by an Indian physicist who is a personal friend of Mehra. None of this could be proved in court, so Kaysen did not pursue the matter further.

Two weeks after this final letter from Kaysen, Mehra happily turned up at my office and stayed overnight at my house. He did not show a trace of nervousness when I took him to eat lunch and supper at the institute cafeteria. Helen must have been there at lunch at the same time but fortunately did not see us. The whole affair is to me completely baffling. There is no possibility that Mehra's discovery of the first Einstein paper was fraudulent. Even his enemies do not deny that the paper is genuine and that Mehra found it honestly. What had he possibly to gain by stealing another well-known paper from Helen's collection? The paper had been published long ago, so that the manuscript was not of historical importance. The thing seems to make no sense.

I have a theory which perhaps does make sense of it. The year 1914 was a crucial year in Einstein's personal life, when his first marriage broke up, and in spite of strong political misgivings he moved from Zürich to Berlin. Helen has often told me that there are personal papers in her collection which she will not allow to be published under any circumstances. It is very likely that such papers might date from 1914. My theory is that Mehra had reason to believe that he could uncover some startling new information about Einstein if he could get a glimpse of some of these personal papers which Helen will not publish. He had only a few seconds to act while Helen's back was turned, so he grabbed into the 1914 drawer and pulled out the wrong paper. Having pulled it out, he was unable to slip it back without being seen. This is an intelligible interpretation of the facts. There are then still two possibilities open. One possibility is that Mehra actually grabbed only the scientific manuscript which Helen reported to Kaysen. The other possibility is that he succeeded in grabbing what he was looking for, in addition to the scientific manuscript, but Helen did not mention the more personal items when reporting the loss to Kaysen. In the second case, Mehra will have kept copies of whatever it was he wanted before sending the package back to Helen. Perhaps someday we shall learn what it was he was after.

The affair is a tragedy for Mehra no matter how it ends. Kaysen has talked about it freely, and it must become common knowledge among historians of science. Mehra's career as a member of the community of scholars is irretrievably ruined. I am very sorry for him.

What he did was crazy, but he had been badly treated. I assume he expected that he could slip the papers out of Helen's drawer, and then on a later visit slip them back in again, without being caught. I wish he had succeeded. But he underrated our Helen. Helen is over seventy, but she is no fool, and she will keep watch over the Einstein papers until her dying breath.

I never succeeded in solving the mystery or straightening out the mess. Mehra continued to be an enemy of Otto Nathan and continued to be my friend. He was never given any further access to the Einstein papers. He continued to work productively as a historian of science. I never found out whether he stole any scandalous document from the Einstein archive. If he did, he carried the secret with him to the grave. He died in 2008. The next letter describes my biggest failure as a psychiatric nurse.

August 5, 1972, Tokyo

In September 1971 there arrived in Princeton a Japanese couple called Taro and Sachiko Asano. Taro was one of the most brilliant of the young Japanese physicists. He had done one outstanding piece of work, and I invited him to Princeton. He came with his bride Sachiko, whom he had married two years before. She is very small. Both of them were quiet and withdrawn. Nobody knew them well. I was the only person to whom Taro came regularly to talk about his work. Sachiko kept herself separated even from the other Japanese people at the institute. They were devoted to each other, and they loved to go touring around America in their blue Ford car. They went on some long trips together.

Taro's work did not go well at Princeton. He failed to repeat his great work of the year before. And I failed to give him the attention he needed. I was busy with one distraction after another. I listened to Taro politely but I did not join in his efforts. I was certainly a disappointment to him. In June and July he got a chronic cough, probably caused by a pollen allergy, and this made him even more depressed.

He stopped coming to the institute and stayed shut up in the flat where they lived. All this time I did not see them and thought they were away touring the West. Sachiko did not tell anybody how worried she was. Imme's mother and nephew were to fly from Germany on Sunday, July 29, and we had to be at Kennedy airport to meet them at five-thirty. On Saturday Sachiko telephoned to tell us that Taro was seriously sick. Imme and I went down to talk to Sachiko. Taro was asleep and we did not see him. Sachiko spoke only about his bronchitis but it was clear that some mental trouble was involved.

At nine on Sunday morning Sachiko called desperately worried, and this time we found Taro awake and could see how bad he was. He talked incoherently and had some kind of persecution mania. He said again and again, "Do not underestimate the power of the high society in Japan." He said that they had hypnotized Sachiko and could make her do anything they wanted. Imme and I brought them to our house and sat in the garden for an hour trying to calm him down. It was clear that we would have to get Taro to return to Japan to find any adequate medical treatment. But there did not seem to be anything useful to do on a Sunday morning. Taro refused vehemently to see a doctor, and we did not pursue the effort to find one. I walked down to the Asano house with them and said good-bye to them there. I said Sachiko should call us if she needed help. I ought to have stayed with them, but it was difficult to sit around all day knowing that Frau Jung was already on her way. Imme wanted me to stay at the Asanos but instead I walked home. A few minutes later Sachiko called to say that Taro had grabbed the car keys and driven off in his car. We rushed down to her and called the police to stop the car on the road. But the police already had reports of a crash on Springdale Road nearby. I went to the scene and found a head-on collision of two cars, with Taro dead behind the wheel of one of them. He must have died instantly, and his face looked more peaceful than I had ever seen it. Luckily the other car was a heavy one and the eight people in it were all alive and on their way to the hospital. I went back to Sachiko and told her of

Taro's death while Imme sat with her crying. We stayed there till we had to drive away to the airport, and some Japanese neighbours took Sachiko to their house.

For the next five days we have been taking turns with the Iitakas and the Okabayashis keeping an eye on Sachiko and packing up her household. Mrs. Iitaka is a marvelous lady who is usually walking around and working with a two-month-old baby nursing at her breast. On Monday Taro was cremated, and on Tuesday there was the ceremony of carrying home the ashes. Imme and I and Mrs. Iitaka went with Sachiko to the crematorium. They brought out the tray of ashes from the furnace, just as it was, untouched by anybody. Then Sachiko came with a little white box and a pair of chopsticks. She carefully picked out of the ashes a selection of small pieces of bones. Then each of us in turn was given chopsticks and made additions to the box. The little white box of bones will be kept in the family shrine forever. Sachiko will not let it out of her sight until she has given it to Taro's mother. The rest of the ashes were poured into a bigger box which is treated much less ceremoniously. It was packed in a suitcase and traveled with Sachiko's heavy luggage. The ashes in the big box will be scattered somewhere in the country that Taro loved, near to his home in Kanazawa.

After the bones were in her care, Sachiko began to get more and more queer. It looked as if Taro's unquiet spirit were haunting her. She drove the Iitakas almost to distraction by not sleeping at night and talking incoherently. We brought two doctors to see her, but she got worse day by day. We had arranged for her to fly alone to Tokyo where her family would meet her this afternoon. But we decided at the last moment this was unsafe. The plane has an hour stop in Anchorage with a complete change of crew, and we could not rely on anybody to take care of her during the stop. So at the last moment I hopped onto the plane with her, carrying only some official papers, a passport, and a toothbrush. The fourteen hours to Tokyo with Sachiko were the most harrowing I have ever spent. She was more crazy than ever. Some of the time crazy in a pathetic way, like Ophelia mourning for Laertes. She would look wildly around the airplane and say over and over, "I want to go back to Japan." Some of the time she was crazy in a sharp way like Hamlet, so that you could not be sure she was unaware of the impression she was making. Some of the time she was as wild as a tiger.

The worst time began with a fight over the official papers which I was carrying to give to Sachiko's father. Among them were four copies of the death certificate which we had gone to some trouble to have issued with the cause of death "auto accident" and no mention of possible suicide. Sachiko grabbed at these papers and began tearing all the death certificates into small pieces and throwing them around the airplane. I unwisely tried to stop her by physical force, which only made her more furious and did not save the certificates. She then stood up with the box of bones in her hands and shouted in a voice to be heard all down the airplane, in strong and fluent English. I had never heard her before talk English so well. "I am Sachiko Asano," she shouted, "and I am taking my husband's ashes back to Japan. And this Professor Dyson who is here is the man who killed him. He killed my husband and he is making plans to kill me too. He is planning to kill me by making me mad just as he did it to my husband. This Professor Dyson and his wife and his children were disturbed by Taro and Sachiko Asano and so they killed us. But they will regret. Professor Dyson's children will die and he will regret. Professor Dyson's wife will die and he will regret." And so she went on and on and on. There was no way to stop her. After she had been through the speech several times in English, she repeated it in Japanese for the benefit of the other passengers. Of course there is a sense in which her accusations are true.

As the hours went by, she became more confused and less vehement. At one point she lay with her head in my lap and slept like a child. After we took off from Anchorage on the second seven-hour hop, she was suddenly playful, fed me by plopping grapes into my mouth, and laughed happily while she made a white beard under my chin with a pillow. Then for a few hours she was lucid and talked about the child she had hoped to have with Taro. Finally as we came close to Tokyo, she became cold and distant. She carefully went through every scrap of paper in her packages and in mine, giving me everything that had the slightest reference to me or to her own medical problems, and taking everything that had reference to Taro. On her package I had written my Princeton telephone number, and she carefully blacked the number over with black ink. Then at the last moment she asked me to write the number back on. Once we had landed in Tokyo she handled all the formalities herself with great competence, including a long argument with the health authority because she had the wrong kind of vaccination certificate.

After that was over we came out into the open, and there was Sachiko's family, at least ten of them, waiting for her. She turned and said to me, "You are a very bad man," then ran to them and was swept up in their embraces. Two of the family came up to me, bowed, and said thank you. I shouted good-bye to Sachiko and walked away, feeling as if the burdens of the whole world had fallen from my shoulders. I think there is a good chance that this volcanic outburst of hatred and malediction will have relieved Sachiko's mind from the unspoken thoughts that were troubling her. I consider it likely that she will now settle down with her family and be her normal self, leaving behind forever the miseries of Princeton. I am glad that I had the final glimpse of her smiling face enveloped in the arms of that big warm Japanese family. I am now flying smoothly home. My charge is done, and I shall enjoy having twenty-four hours of solitary flying to meditate about what it all might mean.

AUGUST 6, 1972, PRINCETON

I came back to Princeton on a bright Sunday morning and found all well at home. But I was wrong in thinking that our part in this story was finished when I said goodbye to Sachiko in Tokyo. Early this morning Sachiko's brother telephoned Imme from Tokyo to apol-

ogize for the fact that they had not had time to thank me properly.
A royal welcome had been prepared for me. They had made frantic
efforts to find me after I walked away. Sachiko was very confused
after she got home. So the happy ending which I had imagined for the
story is not the true ending. Life is never that simple. Now it is time
to see what we can do to help the Liu family, the people who were in
the other car.

September 5, 1972

The aftermath of the Asano tragedy occupied us for some time.
We made friends with one of the Chinese families who were in the
smash and invited the children many times to our pool. They were
finally patched up well enough to go back to their home in Minnesota.
The other Chinese family who were worse injured live in Princeton
and have their own friends here. A few days ago I received a letter
from Sachiko:

> Dear Prof. Dyson, I hope you are all well. While I was in
> Princeton, you were very kind to me in various ways. I wish to
> thank you for your having come to Japan with me. I cannot express
> my thanks for your thoughtful kindness. Many apologies for trou-
> bling you. Trusting that you will forgive me. There was his funeral
> in Kanazawa on August 11. As I lost my husband at a moment, I
> am very sad. I live in Wazima with my parents. I am getting well
> gradually. Now I like to listen the old Japanese song which was
> sung while I was child. Taro respected you very much. He said to
> me you were a great man. When you come to Japan, I would like to
> see you again. Please come to my house with your family. I hope I
> meet you again. Give my kind regards to Imme. I'll write the letter
> to you again. Yours very sincerely, Sachiko Asano.

According to Japanese custom, a widow mourns for her husband for only forty-
nine days. On the fiftieth day, she puts away the mourning clothes, resumes her

maiden name, and returns to her unmarried life. So it was for Sachiko. After a while she remarried, and we lost touch with her. We had no wish to remind her of the tragedy that she had outlived. The robust Japanese acceptance of widowhood is in startling contrast to the Korean tradition. Another of our young physicist friends at the institute died suddenly of natural causes. He happened to be Korean. According to Korean tradition, the widow must stay in mourning clothes for the rest of her life and could never remarry. The widow of our friend could escape this fate only by remaining in the United States.

December 1, 1973, La Jolla

I spent three days at the Salk Institute with the biologists. In that marble palace overlooking the Pacific, I met an Englishman who had just been having a long and moving conversation with Walter Oakeshott. He was visiting Lincoln College in Oxford where Oakeshott is just retiring as warden. Oakeshott was in melancholy mood and talked about his life, telling how both at Winchester and at Lincoln he strove through the years to give the boys or the students some experience of a well-regulated communal life to which they could anchor their intellectual development. His ideal was to make the college something like a big family in which the students and the dons would feel an equal loyalty. But he said, now that he came to retire, he realized the students had never really wanted what he had to offer them. And so he is left wondering what his life's efforts have been worth. I sometimes ask myself the same question in relation to my own children. The only answer is, "Je sème à tout vent" [*I sow to every wind*]. We do our best and hope that in the next generation some seeds will turn out to have fallen on fertile soil. We cannot know which ones.

Walter Oakeshott was a schoolmaster who taught me history at Winchester College. He was then a young man but already an outstanding teacher. He was also well known in the academic world as the discoverer of the manuscript of the Morte d'Arthur *of Sir Thomas Malory, written shortly before William Caxton introduced printing to England in the fifteenth century. Malory's work is the main source in* the English language of the legends of King Arthur and his knights in their mythical castle of Camelot. Some of the legends may have originated from our real king Alfred who lived in Winchester six hundred years earlier. Oakeshott found smudges on the manuscript proving that it had been used by Caxton in his printing shop. The printed version is not identical to the manuscript. Caxton did some editing to the text before he printed it. Oakeshott found the manuscript in 1934, in the library of our school in Winchester, where it had been sitting for 450 years, mislaid because the front page with the title and author's name was missing. After this discovery, Oakeshott was offered an academic position at Oxford, but he preferred to continue teaching at Winchester. He moved to Oxford twenty years later to become warden of Lincoln College. In the short time when I knew him at Winchester, he gave me a firsthand understanding of historical research and a lasting love of history.

Three years after I heard the news about Walter Oakeshott at the Salk Institute, Oakeshott wrote a delightful piece in the Winchester College magazine Trusty Servant with the title, "The Malory Manuscript" (1976). He described how he discovered the manuscript and how it happened to be in the Winchester College library. The first clue to the mystery was the words in the Malory manuscript, "Camelott is otherwyse called Wynchester." Malory and his readers imagined Arthur reigning in Winchester, where Alfred had actually reigned only a little later. The second clue is the coincidence of the date 1485, when Caxton printed the work, with the start of the Tudor dynasty, when Henry Tudor won the battle of Bosworth Field, ending the Wars of the Roses and becoming King Henry VII. Henry had a doubtful claim to the throne and was anxious to establish his legitimacy. He may well have arranged for Caxton to print Malory to strengthen his claim to a royal ancestry. The third and crucial clue is the birth of Henry's firstborn son in September 1486. The son was named Arthur and was born in the bishop's priory in Winchester, just across the street from Winchester College. It was certainly no accident that Henry's queen, Elizabeth of York, the grandmother of the famous Queen Elizabeth who reigned a hundred years later, traveled to Winchester for the birth, and that Arthur was christened with great public ceremony in Winchester Cathedral. The fourth clue is the career of the tutor John Rede who started Arthur's formal education when the prince was five years old. Arthur was a bright student and quickly became fluent in Latin. John Rede was the retired headmaster of Winchester College. Arthur would prob-

*ably have fulfilled his father's hopes and become a capable king, if he had not died
of a fever at the age of fifteen. Oakeshott deduces from these clues that the Malory
manuscript probably came to Winchester at the time of Prince Arthur's birth and
was probably put in the school library by John Rede.*

*Walter Oakeshott was one of those rare people who are equally at home in the
fifteenth century and in the modern world. Winchester College is one of those rare
places where medieval style and beauty are a part of daily life. The man and the
place gave me a long view of history, looking at events with a time scale of centuries rather than years or decades. After this digression to Winchester in the fifteenth
century, I return to La Jolla in the twentieth.*

By coincidence I found my old friend Ted Taylor staying at the
same hotel, and just this week the first article describing his life
appeared in *The New Yorker*. Ted was in a state of great tension, and
we walked and talked for hours up and down the beach. The articles
describe in detail how easy it is to make atomic bombs and how negligent the authorities are in safeguarding the materials. For Ted this
move into the public domain is the culmination of years of efforts to
get the authorities to take the problem seriously. As Ted says, he has
infuriated many of his best friends, and at the worst he may end up
in jail. I was happy to be with him at this turning point of his life. I
admire him now more than ever.

*The New Yorker articles about Ted were written by John McPhee and were
published together in John's book,* The Curve of Binding Energy *(1974). The
book was a best-seller, and Ted did not go to jail. I had also talked for many hours
with John McPhee while he was writing the articles. All three of us were struggling
with the ethical problem, whether it was right or wrong to make public the facts
about homemade nuclear weapons. Were we making homemade weapons less likely
by telling the good guys how to safeguard the materials, or were we making homemade weapons more likely by telling the bad guys how to do it? John McPhee made
the decision to go public, with moral support from Ted and me. Forty years later we
have seen no homemade bombs. Perhaps, after all, John's decision was right.*

21

WHALE WORSHIPPERS
AND MOONCHILDREN

IN 1974 I was working on the theory of adaptive optics. This is the system of control enabling a flexible thin mirror in a big optical telescope to counteract the distortions of the image caused by rapid fluctuations of the atmosphere. The atmospheric turbulence varies on a time scale of milliseconds, and the compensating movements of the mirror must be equally fast. A working adaptive optics system had been developed secretly by the U.S. government for optical imaging of satellites flying overhead. I learned about this secret project as a member of the JASON group of scientists advising the government. Adaptive optics would obviously be of enormous benefit to astronomers peacefully exploring the universe.

Claire Max is a professional astronomer and also a member of JASON. She succeeded in persuading the military to declassify adaptive optics so that we could all work on it openly. Claire continues to be a leader in applying adaptive optics to big telescopes observing faint objects. I solved the problem of designing the best possible computer program that would take the information from the optical image as input and deliver the motions of the flexible mirror as output. I published my solution in a fat paper in the Journal of the American Optical Society. *So far as I know, my solution was never used in a working adaptive optics system. Practical systems that are not optimized work well enough. But it is still helpful to practical designers to know how close they are to the theoretical optimum. The advantages of an optimized system may become greater as telescopes become larger and moving mirrors become more complicated. As a consequence of this jump into astronomical engineering, I was invited to spend the academic year 1974–75 at the Max Planck Institute for Physics and Astrophysics in Munich. We were delighted to spend a year in Germany and renew our contacts with Imme's family. We rented a house in Munich and put the children into German schools.*

In January 1975 my mother died peacefully at the age of ninety-four. She had seven good years after my father's death. So long as he lived, she remained his quiet companion, sitting in his shadow while he enjoyed the limelight. After he died, she came out into the limelight herself, entertaining all of us with her strong opinions and eloquent language. She became once more the forceful personality that she had been as a young woman. Then in her last two years, she was fading and ready to depart. After her death, the letters are to my sister Alice, who remained in our parents' house in Winchester. Alice was three years older than me, had lived most of her life in London, and had returned to Winchester to take care of the parents when she was needed. I was not able to be with her for our mother's funeral celebration.

JANUARY 30, 1975, MUNICH

I wish it shall be a beautiful morning for you on Monday and a ceremony worthy of Mum's demanding spirit. I imagine her floating overhead and inspecting the proceedings with her old sharp eyes now seeing clearly again. Looking at the priest, she would say, "Funny old bird, isn't he?" She always believed that when she died, she would somehow merge her spirit back into the world-spirit from which she came. But I can imagine that after all these years the world-spirit may be finding her a little indigestible. Perhaps she may have to float around for a while before she can merge. Anyhow, when she finally does merge, I am sure she will make her presence felt. On that day the world-spirit will become a bit more demanding and will look at us all with a slightly more critical eye. I hope all this is not just playing with words but in some rudimentary fashion describes the way things really are.

FEBRUARY 16, 1975

Five days in Geneva with the European Southern Observatory, a group of astronomers from six countries who have a telescope in Chile and are the most go-ahead astronomers in Europe. I finally found some people who take my ideas seriously and may follow them up. I gave three lectures, and the room was packed even for the third. After this experience, I decided it makes no sense to go around universities making everywhere the same speech and receiving polite attention. I shall go only where there is a chance of serious activity being stimulated by my coming.

MAY 28, 1975, MUNICH

We did after all go to Westerhausen. All the official formalities were successfully negotiated, and we drove from here to Westerhausen in eight hours with our sleeping bags, dog, and a big box full of presents. We stayed in the big house where Imme lived as a child. It is unbelievable that this big house, which Imme so often described, and which always seemed to me something mythical out of an inaccessible past, is now as solid and familiar to me as my own house in Princeton. It is a magnificent house, the only one of its kind in the village, with fourteen acres of garden and woods belonging to it. It stands there, a little dilapidated but still dominating the landscape. It still belongs to Onkel Bruno. In the old days two families lived in it, Onkel Bruno's family with four children downstairs, Edwin's family with four children upstairs. The eight children grew up together as one big family. The ninth child, Onkel Bruno's youngest, was born after Imme came to the West. Now Onkel Bruno and his wife live upstairs with young Bruno and his wife and baby. The downstairs is rented to another family.

The big house was built by Imme's grandfather who was also the village doctor in Westerhausen. His two sons Bruno and Edwin were both doctors. When the grandfather retired, Bruno took over the practice and Edwin moved to Berlin. Westerhausen is in Sachsen-Anhalt, southeast of Hanover, in the part of Germany that was overrun in 1945 by General Patton's third army. After three months of American occupation, that area was handed over to the Russians in exchange for West Berlin. The Americans moved out and the Russians moved in. Westerhausen became part of the DDR, the German Democratic Republic.

In our honour when we arrived, the whole of Onkel Bruno's family was there, all five of the children who used to be downstairs, four of their wives and husbands, and nine grandchildren. It was a wonderfully warm and friendly atmosphere. There, where the conditions of life are so much harder and the government so much more oppressive, family ties are strong and unbreakable. It was a happy contrast to our spoiled and squabbling relatives in the West. I was particularly struck by the complete trust they have in one another. They all talk freely of the stupidity and corruption of the government, in a house with twenty-six people running around, and without even bothering to shut the windows. I got to know all of them a little and some of them quite well. The men were all working furiously at a wall which they are building around the estate. The old one had collapsed from a storm and long neglect. Every one of the men is in a job where he is frustrated by the system. Not being party people, they cannot hope to get to positions of real responsibility. So they do their jobs as well as the system allows and save their real energies for weekends. They built this wall in Westerhausen with passion. A more solid job I never saw. Beginning with rock foundations three feet deep, then concrete mixed and poured into wooden forms, all hand-made, and on top bricks and mortar.

Coming home across the frontier in the car, our children spontaneously began singing American patriotic songs. I never heard them do that before. That is what the DDR does to you. I am glad they now understand a little better how lucky they are.

AUGUST 1, 1975, LA JOLLA

Tomorrow I fly to Seattle and Emily will fly out to join me there, and together we will go to Vancouver to visit Katrin. I telephoned with Katrin, and she is delighted to have us for the weekend. So we will see her again after five years. Then on Monday we go north to George's territory, to Hanson Island, a little island on the north end of Vancouver Island. George is living there with his new boat and hopes to show us his whales. We shall travel partly by road but mostly by boat.

Last weekend Ken Brower came down here, and I spent two days telling him everything I could think of about George. I like him very much, and we seem to have a lot in common besides our interest in George. He is firmly resolved to go ahead with writing his book. He will come with us to Hanson Island, so he can witness this historic meeting and see me and George together. It is all faintly comic and a bit pathetic, but I shall be glad for practical reasons to have him on the trip. He knows the country and will bring provisions. We shall stay for a week and fly home to Princeton.

Ken Brower is a writer who writes books mostly about nature and people who live in wild places. He got to know George when George was in Berkeley in 1970. After George built his treehouse north of Vancouver and started building beautiful boats, Ken decided to write a book about him. He offered to organize the trip to Hanson Island, where he would play the role of Boswell to George's Johnson. I was glad to cooperate in this enterprise, and my daughter Emily was glad to come with us. I wrote a journal of our trip and sent a copy to my sister.

AUGUST 11, 1975, PRINCETON

The five days on Hanson Island were like a dream. Or rather, it seems that the life up there was reality and everything here is a dream. Most of the time the weather was fresh and fine, as in the Western Isles of Scotland. From our tents in the quiet of the night, we could hear the rhythmic breathing of whales. It is for the whales that George goes to the island. The people on that island are whale worshippers. Their love for these animals has the passionate purity of a religious experience.

August 4 to 8, 1975, Journal of a Visit to the Northern Isles

Monday

Left Vancouver at five-thirty to catch the early ferry to Nanaimo, with Ken Brower and my fourteen-year-old daughter Emily. Ken drove us north along Vancouver Island to Kelsey Bay. Afternoon ferry from Kelsey Bay to Beaver Cove arriving seven-thirty. My son George was at Beaver Cove waiting for us. I had not seen him for three and a half years. Words from somebody's parody of A. E. Housman's "Shropshire Lad" flashed through my head,

What, still alive at twenty-two,
A fine upstanding lad like you.

Because the hour was late and the tide running against us, George did not come in his new six-seater kayak. Instead he came by motorboat with a friend, Will Malloff, who lives on Swanson Island. George had intended to take us to Hanson Island, but Will's boat had engine trouble and so we all stayed overnight at Will's place. This was lucky. We sat up half the night listening to Will's stories.

Will comes from a Doukhobor village and learned the skills of a pioneer from his Russian-speaking parents. He and his wife came to Swanson Island four years ago with two pairs of hands. Now they have a solid and cozy house for themselves, a guest house for their friends, a farm with a Caterpillar tractor, two boats, and a blacksmith's forge with a large assortment of machine tools. Will paid for his two square miles of land by felling and selling a minute fraction of the timber that stood on it. Beyond his homestead, the whole island is untouched forest. The homestead is decorated with woodcarvings done by his wife, the house with wrought iron fashioned by Will himself. The conversation turned to one of my favorite subjects, the colonization of space. I remarked to Will that he and his wife are precisely the people we shall need for homesteading the asteroids. He said, "I don't mind where I go, but I need a place where I can look around at the end of a year and see what I have done."

TUESDAY

Facing Will's homestead, two miles away across Blackfish Sound, stands Paul Spong's house on Hanson Island. Paul also lives alone on his island with his wife and his seven-year-old son Yasha. Paul and Will are as different as any two people could be. Paul is every inch an intellectual. He resigned a professorship at the University of British Columbia to come and live here. His house is a ramshackle affair, made of bits of wood and glass, stuck together haphazardly. One side is covered only with a plastic sheet and leaks abominably when it rains. At the dry end are some beautiful rugs, books, and a 250-year-old violin. We arrived at Paul's place in the morning and found George's kayak at anchor. George had spent the last winter building it, copying the design from the Aleut Indians. He said the Aleuts knew better than anyone else how to travel in these waters. The kayak is blue, covered with animal designs in the Indian style. It has three masts and three sails, rigged like a Chinese junk.

George took us inland to see the tree from which he cut the planks for the kayak. Each plank is thirty-five feet long, straight and smooth and polished. Half of the tree is still there, enough for another boat of the same size. In the afternoon we went out with Yasha in the kayak to look for whales. Since there was no wind and George's crew was inexpert with the paddles, he turned on his outboard motor. I was glad to see that he is no purist. George merely remarked that we must choose either the whales or the motor but not both. We chose the motor and saw the whales only from a distance. At sunset we lay down in the tents which George had prepared for us, on a rocky point overlooking the sea. The evening was still and clear. Soon we could hear the rhythmic breathing of the whales, puff-puff, puff-puff, lulling us to sleep.

WEDNESDAY

It began to rain at midday and continued for about twelve hours. I was glad to taste the life of the pioneers, not only under sunshine and blue skies. George took us out fishing and quickly caught a fifteen-pound red snapper, enough to make a good supper for us all. He spent the afternoon preparing salads and sauces to go with it. The fish itself he baked over Paul's wood-burning stove. During the afternoon Jim Bates arrived with his girlfriend Allison and their seven-month-old baby. Jim is the man who taught George how to build boats. When George was seventeen he worked for a year with Jim building the *D'Sonoqua*, a forty-eight-foot brigantine with living quarters on board for ten people. After she was finished, Jim and George with a group of their friends lived on her for a year, cruising up and down the coast. Then George decided he was old enough to be his own master and quit.

This was my first meeting with Jim. I had already heard much about him from George's letters and expected to encounter another strong, capable pioneer type like Will Malloff. The reality was very different. Jim came up the beach through the pouring rain on crutches. His back is crippled so that he can barely walk. One stormy night last November, he drove the *D'Sonoqua* onto the rocks, close by the Indian village from whose God she takes her name. That night, he says, the God was angry. Allison was with him on board, seven months pregnant. Also with them were two little girls, daughters of Allison. Jim got them all safely to shore, but they lost the ship and everything they possessed on her. Now, nine months later, *D'Sonoqua* is beached not far from Hanson Island, with gaping holes in her bottom, her inside furnishings rotted and wrecked. Jim has not given her up. Every spare minute he drags himself to work on her and dreams of getting her afloat. He is skipper of the *D'Sonoqua* still.

I looked into the eyes of this noble wreck of a man, and I saw the true image of Captain Ahab. With his wild, far-away eyes, obsessed with impossible visions, he is destined to drive himself and all those near him to destruction. And I looked into Allison's eyes, full of patience and gentleness, and saw in them the unconquerable loyalty which will never allow her to abandon this monomaniac to his fate. It was pitch dark when Jim and Allison left. I watched them walk slowly down the beach to the boat, in the dark and pouring rain, Jim on his crutches, Allison carrying the baby in her arms. It was like the last act of *King Lear*, when the crazy old king and his faithful daughter Cordelia are led away to their doom, and Lear says,

Upon such sacrifices, my Cordelia,
The gods themselves throw incense.

Tragedy is no stranger to these islands.

THURSDAY

In the morning it was still raining. Emily and I lay comfortably in our tents while George gave an exhibition of his skill as an outdoorsman. In an open fireplace under the pouring rain, using only wet wood from the forest, a knife, and a single match, he lit a fire and cooked pancakes for our breakfast. In the afternoon the sun came out, and we went for a longer ride in the kayak. This time there was some wind, and we could try out the sails. She sailed well downwind, but without a keel she could make no headway upwind. George had made a pair of hydrofoils, which will be fixed to her sides as outriggers and will give her enough grip on the water to sail upwind. But it will take him another month to make the outriggers and put the whole thing together. In the meantime, we have been improving our skill with the paddles.

Since Thursday was our last evening on the island, we went to visit with Paul and his family. When it was almost dark, the whales began to sing. Paul had put hydrophones in the water and connected them to speakers in his house. The singing began quietly and grew louder and louder as the whales came close to shore. Then the whole household exploded in a sudden frenzy. Paul grabbed his flute, rushed out onto a tree trunk overhanging the water, and began playing weird melodies under the stars. Little Yasha ran beside Paul and punctuated his melodies with high-pitched yelps. And louder and louder came the answering chorus of whale voices from the open door of the house. George took Emily out in a small canoe to see the whales from close at hand. They sat in the canoe a short distance from shore, and George began to play his flute. The whales came close to them, stopping about thirty feet away, as if they enjoyed the music but did not wish to upset the canoe. So the concert continued for about half an hour. Afterwards we counted the whales swimming back to the open sea, about fiteen in all. They are of the species popularly known as killer whales, but Paul calls them only by their official name, orca.

Anybody who witnesses this ceremony of the whales cannot fail to be profoundly moved by it. And indeed it is Paul's purpose to use the whale songs to awaken the conscience of mankind, so that the whales may be preserved from extinction. The ceremony itself has a religious, mystical quality. But one may see and hear it, and be moved by it, without believing that whales are gifted with superhuman intelligence or supernatural powers. The ceremony is a natural one and seems strange only because we are unaccustomed to the idea of a whale behaving like a dog or a horse. Paul considers it an established fact that the whales enjoy and respond to his flute playing. How much more they may understand, he does not pretend to know.

FRIDAY

Our last day. It happened to be shortly after new moon, so that the tides were stronger than usual. We woke early to find the sun shining, sat on our rock overlooking the water, and watched the morning birds. Kingfishers skimming below our feet, eagles soaring above our heads. Between Hanson and Swanson Islands, about a mile from shore, there is a strong tide race. That morning it was fierce, making a white streak on the blue sea. By and by we saw a little black speck move into the white area and heard the distant putt-putt of a motor. George saw more than Emily and I did. He said quietly, "Those people have some nerve, going with an open boat into that kind of water." A few seconds after he spoke, the black speck disappeared and the noise stopped. George at once moved into action. Taking Ken with him, he ran to Paul's motorboat, an unsinkable affair made of rubber, and within two minutes was on his way out. From the shore we could see nothing for the next half-hour. I roused the Spongs and helped them heat up their stove. Then the rubber boat reappeared, and we could make out four figures in it. They came ashore, and I helped the old man stagger up the beach, his hand in mine as cold as ice. We wrapped them in blankets and sat them down by the stove.

An old man and a young man, both loggers on strike, had decided to go out with their aluminum boat to dig clams. It was a lovely morning, clear and still. They never imagined that one could capsize on such a morning. Luckily they had had the sense to cling to their capsized boat and not try to swim to shore. But George said they were close to the end when he found them. The old man had not been able to move his arms or legs anymore. In that icy water nobody can last long. While they revived, George cooked hot tea and pancakes on the stove. Then he radioed to their families to send a boat to take them home. The old man afterwards told me how it had felt. He said he knew his life was over and he was ready to go under. When the rubber boat appeared he thought he was seeing visions. Only when Ken and George hauled him aboard did he believe it was real. In the afternoon he and I chatted again over cups of tea. He turned out to be intelligent and well-read, and he asked me many questions about my life and work at Princeton. And I said, "But it seems to me now the best thing I ever did in Princeton was to raise that boy."

Toward evening a big solid tugboat arrived to take the two loggers home. In the meantime George and Ken had rescued their boat and beached it on Swanson Island, taken their motor apart, and soaked the insides in fresh water. So the loggers went home with their boat and their motor intact, ready for another day. It was now time for us also to depart. George took us in the rubber boat to catch the night ferry going south from Beaver Cove. He was apologetic because we went home empty-handed. He had intended to spend the last day with us salmon fishing, so that we could take with us two big salmon, one for his friends in Vancouver and one for my family in Princeton. I said to him, "You don't need to apologize. Today you went fishing for something bigger than salmon." And that was our good-bye.

Ken Brower's book The Starship and the Canoe *(1978) turned out to be a double biography, half about George and half about me. It told a slightly romanticized story of both of us. It was helpful to George, making him a local celebrity in British Columbia. It was helpful to me, bringing me new friends in the environmental community. It frequently happened that strangers would write to me thanking me for the book, under the impression that I was the author. I hastened to write back telling them that it was actually written by Ken Brower. Looking at the book now almost forty years later, I am grateful to Ken for providing an accurate and perceptive account of the father-and-son drama through which George and I lived, beginning with anger and rebellion, ending with pride and joy.*

My portrayal of Jim Bates as a tragic Captain Ahab turned out to be totally wrong. After our visit, Jim slowly recovered from his injuries and succeeded in patching the holes in the bottom of D'Sonoqua. *To my astonishment, he made her seaworthy, turned her upright, and got her afloat. He continued for several years to sail her up and down the coast of British Columbia. Then he sailed her down through the Panama Canal to the Atlantic. The last time George saw her, he was on St. Simon's Island in Georgia and unexpectedly ran into Jim Bates at the island's grocery store.* D'Sonoqua *was anchored in the nearby river, making her*

way up the Atlantic Coast. Jim eventually sailed her to the Bay of Fundy, where
she now still sits, up on land on a property where Jim rents cabins to tourists.

I also misjudged Will and Georgeanna Malloff when I described them as a suit-
able team for colonizing an asteroid. Less than a year after our visit, Georgeanna
abandoned Will and left the island. She is an artist and needs a community to
give meaning to her life. Five years of solitude was enough. Will stayed on the
island, unwilling to move but unhappy to be alone. He died in 2015 at Alert Bay.

The battle of Princeton was a turning point in the American War of Inde-
pendence. It was the first major battle that Washington won, proving to doubtful
fence-sitters in America and to doubtful observers in Europe that the rebellious
colonists might actually defeat the British Empire. Princeton is justly proud of this
victory, paid for by the citizens with heavy losses of life and property.

January 10, 1977

The big event this week was the Battle. January 3, 1977, was
exactly two hundred years from the original battle, and they reen-
acted the whole thing as precisely as they could. Five hundred real
redcoats from England came over for the occasion, and a suitably
rugged and miscellaneous crowd of Americans were collected from
various places to make up Washington's army. The only difference
from the original battle was that nobody was hurt. About ten thousand Princetonians came to watch and got in the way to some extent.
The great good luck was that the weather was the same as on January
3, 1777, a cold sunny day with about three inches of snow on the battlefield and the ground hard frozen. So not only Washington's troops
but also the spectators were able to move around without producing a
sea of mud. The battlefield was still clean and white when it was over.
What people forget is that the real battle destroyed about half of the
houses in Princeton, and more people died afterwards from cold and
sickness than died in the battle.

Our home is on Battle Road, on the edge of the Battlefield Memorial Park, a
protected area of grassland where the reenactment of the battle took place. Histo-

rians are still disputing how much of the fighting in 1777 was actually within the park. Much of the battle was probably in the built-up area of the town, where many buildings were destroyed.

JANUARY 29, 1977

Today I went to the first meeting of a citizens' committee which is supposed to decide for the town of Princeton whether the biologists at the university are to be permitted to work with recombinant DNA. I was asked to serve on the committee and agreed to do so because this is an important question and I should not stand aside. The committee will involve a great deal of work, and is supposed to produce a final report by May 1. We were told to expect to put into it about ten hours of work per week for ten weeks. It will probably add up to more than that.

The modern era of genetic science began in 1976 with the discovery of recombinant DNA, popularly known as gene splicing. Recombinant DNA is the genetic hybrid produced by splicing a gene from one species into another. Recombinant DNA allows an experimenter to take genes from a bacterium or from a mouse and insert them into a living human embryo. Experimenters could break the barrier that nature had put between species. Biologists all over the world understood that such experiments could raise serious problems of ethics. An international meeting of biologists at Asilomar in California agreed to impose a set of guidelines, deciding which experiments should be allowed and which should be forbidden. The biologists at Princeton University accepted the Asilomar guidelines, but the municipal authorities of Princeton Township were unsure whether the Asilomar guidelines were sufficient to ensure public safety. The municipal authorities set up our committee to find out whether recombinant DNA experiments were acceptable to the Princeton community. Our job was to educate ourselves and also to educate the public concerning the possible costs and benefits of the new technology. Most of our time was spent listening rather than talking. We invited expert and nonexpert witnesses to come and express their opinions. Anyone who wished to be heard was heard. We treated every witness with respect, no matter how long-winded or igno-

rant they might be. As a result of our patience, the public debate remained friendly and nobody felt excluded. When we finally announced our conclusions, they were generally accepted as fair and reasonable.

Today we had our first meeting, and I found to my surprise that the main subject of discussion was whether I am fit to serve. A lady who is a vociferous opponent of DNA research challenged my membership on the ground that I have taken a public position in favor of the research and cannot have an open mind. So we argued about this for the whole afternoon. I said I would be happy to step down if the township authorities ask me to. I really hope they will. But I feel an obligation to stick it through if they decide to keep me on. This will be decided next week.

The lady who opposed my membership was Susanna Waterman, an environmentalist who sincerely believed that DNA research was a violation of nature. The township denied her objection, and I continued to serve on the committee. Quite soon Susanna and I became friends, and I valued her presence as the most thoughtful and eloquent voice on the committee. In the end, I voted with the majority to allow DNA experiments, and Susanna voted with the minority to forbid experiments. But we learned to respect each other's opinions and stayed friends.

MARCH 8, 1977

Our Princeton committee has been meeting twice a week for five weeks, and we are getting to know each other very well. I am already quite sentimental about the group and shall be sorry when our task is done. We shall be friends when it is over, however much we may disagree about details. I discovered that Emma Epps, the black woman who is one of the most thoughtful and sensible of the group, lived for thirty years as a maid in the house two doors away from ours. Dora Panofsky used to talk in those days about her wonderful Emma, but we never got to know her as a person. Now she is one of the leaders of the black community and a powerful person in Princeton. If only

Robert Armstrong and his friends in Rhodesia would understand that black power is not the end of everything.

Dora Panofsky was the wife of the famous art historian Erwin Panofsky and herself also an art historian. They had twin sons who both became famous scientists. Emma Epps ran the household and brought up the twins. One of them was Wolfgang Panofsky, a physicist whom we knew well. He spoke of Emma Epps with affection and respect. I felt the same way after getting to know her on the committee. Robert Armstrong was a cousin of mine who settled in Rhodesia, owned a tea plantation there, and stayed there through the years when the white supremacy government of Ian Smith ran the country and afterwards when the black power government of Robert Mugabe changed its name to Zimbabwe. After the country became Zimbabwe, I wrote to my cousin asking how he was getting along with black power. He replied, "If we had not known how to get along with the blacks, we would not have lasted six weeks." He continued to live peacefully in Zimbabwe until he died many years later.

APRIL 14, 1977

Our committee had its crucial meeting last Monday when we each had to decide on the main question, whether to approve the plans of the university. The vote was seven to two in favor with two people absent, and one of the absentees has told us he also votes yes. I am much relieved. It now remains for us to find out if we can agree on the wording of our report.

The final vote was eight to three. The minority consisted of Susanna Waterman, Emma Epps, and Wallace Alston who was pastor of the Presbyterian church. The committee remained divided, because the majority interpreted our task to be to judge whether the DNA research would be an immediate danger to public health, while the minority wanted to judge whether the research might be a violation of long-range ethical principles. Though I voted with the majority, I felt more personal sympathy with the minority, and I was delighted to see that the minority opinions were heard and understood by the public. We quickly decided that we could not write a unanimous report as requested by the township. The majority and minority wrote separate reports, so that the township authorities could hear the arguments of both sides. After considering our two reports, the township finally decided to accept the advice of the majority and allow the university to go ahead with gene-splicing experiments. In forty years since this decision was made, no public health hazards have arisen from the experiments.

FEBRUARY 8, 1978, VANCOUVER

Here I am in Vancouver taking a long walk on the beach with George. I never heard him talk so much. All about his expedition of last summer. At least enough material to fill a book. The expedition was completely crazy, with twelve people who had never been together before, most of them with no experience of the ocean, wandering around the Pacific with six little boats. They started quarreling immediately, some of them panicked, some of them were lost for weeks at a time, and some of them ran out of food. The amazing thing is that George got them all back alive with no loss of boats or equipment. He said he has no desire ever to lead an expedition again. But obviously he has a gift for it and I think this will not be his last. He said he made more real friends and more enemies in that summer than he ever had before. The friends will last and the enemies he will not see again. I found George less strange than he used to be. He is talking more plainly and more freely. He is self-confident in his plans for the future. He will not build any more boats for the time being, but will run a school where people can learn to build their own boats and sail them. He has already leased a big empty house in the woods near to his treehouse and will begin his school there. The house is now derelict, but for George it is no problem to fix it up. The big surprise, George owns a car and drives around in it. He already drove it to California and back. It is a twelve-year-old Volvo, and he bought it derelict for two hundred dollars. He took it apart and put it together so that it now goes beautifully.

Lastly, Katrin. This Katrin who for so many years was living her crazy life, and we never knew where she was. Now here she is, working as a secretary in the very same building where I am staying, the faculty club of the university. This evening we had a faculty dinner, and she was invited to join us. All the professors know her as a secretary but were astonished when she appeared as my daughter. She came looking very elegant in a long black dress and a necklace that Imme bought for her in Princeton. The necklace matches her eyes and is exactly right.

FEBRUARY 12, 1978, VANCOUVER

I spent two days at George's hideout in the woods. Just Katrin and George and I. Sunshine all day long. Sea birds and snowy mountains. On Saturday Jason came out to join us for supper, and George cooked an immense brew of lasagne with mushrooms that he had picked in the mountains. I don't worry about anything when George is in charge, not even mushrooms. At night I slept in the house which he has furnished with two big old wood stoves, one for cooking and one for heating. Like a true seaman, he makes everything clean and neat. At bedtime I looked out of my window into the starry night and watched him climb like a squirrel up to his treehouse. I could hear the click of his door opening and shutting, and then I could see the golden light of his oil lamp shining on his wooden ceiling. I was looking up through his window into his house at an impossibly high angle. That little golden lighted window seemed to be floating among the stars.

In Vancouver one of the physics professors gave a supper party to which George, Katrin, and Jason were invited. It was good to see the two cultures sitting together around the supper table. After supper George gave his slide show of the 1977 expedition. It was a magnificent show, and George told his stories well. He is quite at his ease now in any sort of company. He has the air of a man who knows where he is going. George's plans for the future are definite and ambitious. He wants to buy a particular island near to the place we camped in 1975.

This island would be the permanent base of his Baidarka School. Until the money materializes he will keep the school at the Indian Arm base where he has his workshop. To run the school at Indian Arm he will need about $20,000 a year. He has budgets and plans written down for prospective donors. He told me the main thing he has learned in this year of the 1977 expedition has been how to ask for money. The 1977 expedition cost $11,000, which he successfully raised and spent. The final accounting showed he had overspent the account by exactly ten dollars. The two weapons he has in raising money are his slide show and his air of self-confidence. This side of George, the administrator, the man who can handle money, was absolutely new to me. He said it is new to him too. And yet after all, it is obvious where it comes from. He begins more and more to resemble his grandfather.

Carl Sagan and Edward Wilson were first-rate scientists who knew how to communicate science to the public. Both aroused some hostility, Carl because he was too successful as a television star, Edward because his views on human biology were politically incorrect. In 1977 Carl had given the Christmas Lectures for Children at the Royal Institution in London, a famous lecture series started by Michael Faraday in 1827.

FEBRUARY 18, 1978, PRINCETON

I ran into Carl Sagan in Washington and told him you had enjoyed his talks. He said he had a very good time with the Christmas lectures and found the children delightful, except for the two royal princes who were there and made everyone feel stiff and uncomfortable. He said these princes are so well trained, they can talk intelligently about everything and are interested in nothing. Back from the London lectures, he was immediately invited for a family evening at the Carters, with Jimmy, Rosalynn, Amy, Jeff, and Caron, chatting about Mars and Jupiter and extraterrestrial intelligence. He found Jimmy not so well trained as the English princes but interested in everything.

The trip back from Washington was more exciting than usual.

We hit a bad patch where the freezing and thawing of winter made the track bumpy. After a few violent bumps the whole train jumped off the rails. We ended up tilted half over but not quite overturned. Nineteen people were injured, none seriously. After a long wait we were rescued by buses which took us back to Union Station in Washington. I was sitting in the waiting room, battered and disheveled, when I noticed Ed Wilson, a friend of mine who is a professor of biology at Harvard, sitting in the next chair, looking more battered and disheveled than I was. He had a pair of crutches, one leg in a cast, his hair dripping wet, and his clothes looked as if he had been in a fight. I asked him if he had also been in the train wreck and he said no. But he had a bad week. First he slipped on some ice in Boston and broke his leg. But he had agreed to talk to the AAAS (American Association for the Advancement of Science) meeting where I also was talking and painfully dragged himself on his crutches to Washington for the meeting. He is the world's greatest authority on the behavior of ants, and he has written a book, *Sociobiology* [1975], pointing out the analogies and differences between social behavior in insects and human beings. This is politically dangerous ground, and the young radicals at Harvard have been accusing him of being a fascist and a racist. When he came to give his talk in Washington, a bunch of young hooligans invaded the platform, grabbed the microphone out of his hands, told the audience what an evil character he is, and ended by emptying the speaker's water jug over his head. After this he gave his talk, and the audience gave him a standing ovation. But he said that is the last time he ever comes to a AAAS meeting.

April 19, 1978, Princeton

The main event of the last weeks has been the *Moonchildren* affair, which has been in all the local newspapers and even for a few days in the *New York Times*. *Moonchildren* is a play which Imme and I saw in New York, a good play, rather Chekhovian in theme, about a group of students in a rooming house during the Vietnam years.

Miriam's high school drama class decided to perform it at the high school, and Miriam has beeen working at the sets and the production with all her heart and soul. She also had a minor part as Aunt Stella. Then just two weeks before opening night, the high school principal announced that the play must be expurgated and certain four-letter words omitted. The children got a lawyer and took the case to court in Trenton as a violation of their rights of freedom of speech. For two days Miriam was at the trial in Trenton, learning a lot more about life than she would have learnt in class at the high school. The children lost the case. But the production, even without the four-letter words, was a huge success and crowds of people had to be turned away from the door.

After the expurgated Moonchildren *production at the high school was over, the Unitarian Church invited the children to put on an unexpurgated production at the church meeting room.*

APRIL 27, 1978

We went to the unexpurgated version of *Moonchildren* at the Unitarian church. The acting was magnificent. All this fuss has made the children do a far better job than they would have been capable of under normal circumstances. Unfortunately the drama teacher Arlene Sinding who produced the play will now be leaving the school. She is not exactly kicked out, but after this affair it would be hard for her to stay. This morning Miriam was working hard in the bathroom getting the makeup out of her face and hair.

From the *Moonchildren* program, April 1978, by Miriam Dyson: To describe the sixties in one paragraph would be to describe the color red to someone who is blind. One could ramble on about Vietnam, the riots, the generation gap, any of these things which gave the sixties its label. But the color would be left out. Color is a feeling. The sixties is a feeling, a color. A color seen only by those who lived, fought, loved and survived the sixties.

June 15, 1978, Princeton

On Saturday morning I took the train to Washington and walked as usual the two miles through the ghetto area to my hotel. People friendly as usual. No trouble. The hotel is in the posh area two blocks from the White House and four blocks from the academy. After lunch I walked out of the hotel to go to a meeting at the academy. A young black man came up to me and began talking in an educated voice about how to find the quickest way to the Kennedy Performing Arts Center. I walked along with him and chatted in a friendly way. His conversation seemed a bit strange, he used a lot of big words that he didn't understand, but I was quite unsuspecting. Then as we walked past some bushes his accomplice came quietly behind me and grabbed me around the neck. The accomplice must have been hiding in the bushes but I never saw him. They quickly pulled me into the bushes, hit me three times on the side of the head, and left me lying on the ground while they picked my pockets and briefcase. All they got was a wallet with seventy-five dollars and some pictures of the girls.

I was fully conscious the whole time and in a state of spiritual peace that is good to remember. I saw the bright sunshine filtering down through the bushes, and it was beautiful. I thought, very likely these fellows will put a bullet into me to keep me from talking. I was quite ready for death, and it did not seem frightening at all. I thought, life has been good to me and this death is also good, with the bright sun and the green bushes. It is good to know that death can be so friendly. Then after a few seconds the men ran away and I picked myself up and found a Good Samaritan who drove me the short distance to the academy. My entrance into the academy was quite dramatic. There I was among friends. Best of all, the thugs didn't even break my glasses. I had the bifocals on when they attacked me. At the hospital in Washington the police came to talk to me, and I told them exactly where the attack occurred and asked if they could look for the glasses. An hour later a grinning policeman came in with the bifocals, not even scratched. Fortunately they let me out of the hospital right away, and so I traveled back to Princeton in great style in a private air-taxi, landing at the little airstrip only two miles from home. In less than five hours from the time of the accident, Imme and the girls came to get me. I looked so ugly that first evening that Miriam couldn't bear to look at me. But now I am back to my normal beauty with some glorious sunset colors added. I had only two restless nights and no bad pain. For the first two days I was seeing double. That was all. No concussion, no headaches. Amazing luck.

The whole of Monday I was pushed around from one doctor to another, and my head was shot through with X-rays. The score was three fractures, upper jaw, right cheekbone, and the floor of the right eye. This spoilt our family record of not having any broken bones. This morning I passed another important milestone. I sneezed for the first time since the accident, and none of the loose bones fell clattering to the floor. The amazing thing is how well everything functions, brains, eyes, ears, teeth all okay. I only missed one working day at the Institute. Today I called the plastic surgeon, and he said he doesn't want to see me anymore. Thank God. Now I am out of his clutches. When Imme and I went to see him on Monday, he was itching to get me laid out on the table so he could begin to carve. I must confess he scared me more than the thugs who beat me up in Washington. I am now free of him and can let Nature do her job of healing.

A close encounter with death teaches us important truths about human nature. We are not only social animals. We are also fighting animals. We may dream of universal brotherhood, but when the bugle sounds, we run bravely into battle. Battered and bruised in a surprise attack, I found myself unexpectedly reacting to it with calm courage and joy. I could handle it much better than I would ever have imagined. In every culture and every battlefield, from the Spartans at Thermopylae to the Jews at Masada, the men who died in battle are remembered and honored as heroes. In the battle of Princeton, George Washington rode his horse at the head of his troops, a conspicuous target for the British sharpshooters. He knew that an act of reckless bravery would make him a more effective leader of his country in the long struggle that lay ahead. In the future as in the past, reckless bravery will be honored, and fighters will be leaders. We must try as hard as we can to make peace with our enemies and get rid of weapons of mass destruction, but we cannot expect to extinguish the fighting spirit and tribal loyalty that are deeply ingrained in our nature. Perpetual peace is a worthy goal, but it is likely to remain out of our reach.

A world of turmoil and violence is our legacy to future generations. They need to understand why science has failed to give us fair shares and social justice, and they need to work out practical remedies. This is not a job for scientists to do alone. It will need a worldwide collaboration of scientists with economists, political activists, environmentalists, and religious leaders, to lift science and society out of the swamp where we are stuck. Pure science is best driven by intellectual curiosity, but applied science needs also to be driven by ethics. Our grandchildren will have a chance to make this happen. One of them is the girl with the pink Afro who spoke at the beginning of this book.

BIBLIOGRAPHY

ARNOLD, MATTHEW. "The Forsaken Merman." In *The Strayed Reveller and Other Poems*. London, 1849.

BERNSTEIN, JEREMY. *The Analytical Engine: Computers, Past, Present and Future*. New York: Random House, 1964.

BLAKE, WILLIAM. "Eternity." In *The Portable Blake*. Edited by Alfred Kazin. New York, Viking Press, 1946.

BROWER, KENNETH. *The Starship and the Canoe*. New York: Harper and Row, 1978.

CHADWICK, JOHN. *The Decipherment of Linear B*. New York: Cambridge University Press, 1958.

CLARK, RONALD. *Einstein: The Life and Times*. New York: World, 1971.

CLARKE, ARTHUR C. *Childhood's End*. New York: Ballantine, 1953.

———. *Profiles of the Future: An Inquiry into the Limits of the Possible*. New York: Holt, Rinehart and Wilson, 1962.

CLOSE, FRANK. *Half-life: The Divided Life of Bruno Pontecorvo, Physicist or Spy*. New York: Basic Books, 2015.

DAY LEWIS, CECIL. "Birthday Poem for Thomas Hardy." In *Poems, 1943–1947*. London: Jonathan Cape, 1948.

DYSON, FREEMAN J. "Death of a Project." *Science* 149 (1965): 141–44.

———. *Imagined Worlds*. Cambridge, Mass.: Harvard University Press, 1997.

———. "A New Symmetry of Partitions." *Journal of Combinatorial Theory* 7 (1969): 56–61.

———. "Search for Artificial Stellar Sources of Infra-red Radiation." *Science* 131 (1960): 1667–68.

———. "Some Guesses in the Theory of Partitions." *Eureka* 8 (1944): 10–15.

———. "Tomonaga, Schwinger and Feynman Awarded Nobel Prize for Physics." *Science* 150 (1965): 588–89.

———. *Weapons and Hope.* New York: Harper and Row, 1984.

DYSON, GEORGE. *Grenade Warfare: Notes on the Training and Organization of Grenadiers.* London: Sifton Praed and Co., 1915.

DYSON, GEORGE B. *Project Orion: The True Story of the Atomic Spaceship.* New York: Henry Holt, 2002.

DYSON, VERENA HUBER. "Thoughts on the Occasion of Georg Kreisel's Birthday." In *Kreiseliana: About and Around Georg Kreisel.* Edited by Piergiorgio Odifreddi. Wellesley, Mass.: A.K. Peters, 1996.

EHRENBURG, ILYA. *The Thaw.* Translated by Manya Harari. Chicago: Henry Regnery, 1955.

EINSTEIN, ALBERT. *Albert Einstein–Max Born Briefwechsel, 1916–1955.* Munich: Nymphenburger Verlagshandlung, 1969.

ETZIONI, AMITAI. *The Hard Way to Peace.* New York: Collier Books, 1962.

FEYNMAN, RICHARD P. *What Do You Care What Other People Think?* New York: W. W. Norton, 1988.

HARDY, GODFREY H. *A Mathematician's Apology.* Cambridge, U.K.: Cambridge University Press, 1940.

HERBERT, GEORGE. "The Pulley." (1633). In *The Poems of George Herbert.* Oxford: Oxford University Press, 1907.

HILLIS, DANIEL. "Intelligence as an Emergent Behavior: or, the Songs of Eden." *Daedalus* 117 (1988): 177.

HOFFMANN, BANESH, AND HELEN DUKAS. *Albert Einstein, Creator and Rebel.* New York: Viking Press, 1972.

HUXLEY, ALDOUS. *Brave New World.* London: Chatto and Windus, 1932.

———. *The Doors of Perception.* New York: Harper & Brothers, 1954.

IBSEN, HENRIK. 1879. *A Doll's House and Other Plays.* New York: Modern Library, 1950.

INFELD, LEOPOLD. *Quest: An Autobiography.* 1941; Reprint by American Mathematical Society, 2006.

KENNAN, GEORGE F. *Soviet-American Relations, 1917–1920.* Princeton: Princeton University Press, 1965.

KEYNES, JOHN M. "Newton, the Man." (1943). In *Newton Tercentenary Celebrations.* Cambridge: Cambridge University Press, 1946.

LAWRENCE, DAVID H. *Sons and Lovers.* London: Gerald Duckworth, 1913.

McPhee, John. *The Curve of Binding Energy*. New York: Farrar, Straus and Giroux, 1974.

Mehta, Madan Lal. *Random Matrices*, 3rd ed. Amsterdam: Elsevier, 1967.

Mehta, Ved. *Face to Face*. Boston: Little, Brown, 1957.

———. *Walking the Indian Streets*. Boston: Little, Brown, 1960.

Nathan, Otto, and Heinz Norden. *Einstein on Peace*. New York: Simon and Schuster, 1960.

Oakeshott, Walter. "The Malory Manuscript." *Trusty Servant* 41 (May 1976).

Pour-El, Marian B., and Ian Richards. "The Wave-Equation with Computable Initial Data Such That Its Unique Solution Is Not Computable." *Advances in Mathematics* 39 (1981): 215–39.

Prescott, William H. *The History of the Conquest of Mexico*. New York: Harper and Brothers, 1843.

———. *A History of the Conquest of Peru*. New York: Harper and Brothers, 1847.

Pupin, Michael. *From Immigrant to Inventor*. New York: Charles Scribner's Sons, 1922.

Rumpf, Hans. *Der hochrote Hahn*. Darmstadt: Mittler, 1952.

Spicer, Paul. *Sir George Dyson, His Life and Music*. Woodbridge: Boydell Press, 2014.

Thackeray, William M. (pseud. M. A. Titmarsh) *The Rose and the Ring: A Fire-side Pantomime for Great and Small Children*. 1834; reprint by London: Macmillan, 1934.

Tolstoy, Lev N. *War and Peace*. 1869; reprint by Moscow: OGIZ, Collection of Government Publications, 1941.

Wilson, Edward O. *Sociobiology: The New Synthesis*. Cambridge, Mass.: Harvard University Press, 1975.

Yeats, William B. "Aedh Wishes for the Cloths of Heaven." In *The Wind Among the Reeds*. London: John Lane, Bodley Head, 1899.

INDEX